THE IMPLEMENTATION OF PROLOG

The present translated volume has been revised and updated since its original publication in French. All the programs developed in this book are available by anonymous ftp at ftp.cnam.fr in the directory pub/ Boizumault. We thank Mr. Bortzmeyer and Le Conservatoire des Arts et Métiers de Paris for their help.

THE IMPLEMENTATION OF PROLOG

Patrice Boizumault

Co-translated by Ara M. Djamboulian and Jamal Fattouh

PRINCETON UNIVERSITY PRESS
PRINCETON, NEW JERSEY

Published by Princeton University Press, 41 William Street, Princeton,
New Jersey 08540
In the United Kingdom: Princeton University Press, Chichester, West Sussex
This book was originally published in French as *Prolog: L'Implantation*,
copyright © 1988 by Masson Editeur, Paris

Library of Congress Cataloging-in-Publication Data

Boizumault, Patrice, 1959–

     The implementation of Prolog / Patrice Boizumault.

     p.   cm. — (Princeton series in computer science)
     Includes bibliographical references (p.   ) and index.
     ISBN 0-691-08757-1
     1. Prolog (Computer program language) I. Title. II. Series.
QA76.73.p76B65   1993
005.13'3—dc20   92-25951

Princeton University Press books are printed on acid-free paper and meet the
guidelines for permanence and durability of the Committee on Production
Guidelines for Book Longevity of the Council on Library Resources

Printed in the United States of America

10 9 8 7 6 5 4 3 2 1

To Dominique
To Roger and Odette

CONTENTS

PART I
FUNDAMENTAL PRINCIPLES OF THE LANGUAGE

C H A P T E R 1
UNIFICATION

C H A P T E R 2
RESOLUTION AND PROLOG CONTROL

C H A P T E R 3
IMPROVING PROLOG CONTROL

PART II
PRINCIPLES AND TECHNIQUES OF IMPLEMENTATION

CHAPTER 4
CONTROL AND STACK(S) MANAGEMENT

CHAPTER 5
REPRESENTATION OF TERMS

CHAPTER 6
DETERMINISTIC CALL RETURNS

CHAPTER 7
LAST-CALL OPTIMIZATION

CHAPTER 8
CLAUSE INDEXING

CHAPTER 9
COMPILATION OF PROLOG

PART III
IMPLEMENTATIONS

FOREWORD

1972 was the official birthdate of Prolog, recognized as the first operational implementation of LOGic PROgramming. A few years were sufficient to establish Prolog as a respectable and recognized language. The long list of languages that borrow its name in part or in whole testifies to the importance of the Prolog phenomenon.

The history of programming languages is rich in examples where the specifications of a language and its implementation exist before its semantics are defined. This is not the case with Prolog, which benefited from the beginning from a well-defined semantics. This semantics evolved through successive refinements, unlike so many languages often composed of awkward patches whose legitimacy remains obscure to the user.

Contrary to a well-established opinion, Prolog is not an "easy" language, even if programmers are initially seduced by its apparent simplicity. Its basic mechanisms, unification, and nondeterminism are more difficult to grasp than an inference principle. As with Lisp, one cannot reasonably claim to be an accomplished programmer in Prolog without having a profound understanding of its implementation. Patrice Boizumault's book comforts the programmer puzzled by the subtlety of the tool mechanisms.

Fortunately, it is now a trend both that the apprenticeship of Lisp includes an exercise in writing a Lisp interpreter and that the teaching of Prolog includes the implementation of a Prolog interpreter in Lisp, C, or Prolog itself. It is not just another routine stylistic exercise: on the contrary, its purpose is to provide the programmer with an in-depth knowledge of the sophisticated mechanisms he or she is using. Today's students are thus well-prepared to reach the bottom line, and we can only advise older programmers in the same manner. This book is the bread and butter for this provided programmers have adequate intellectual curiosity, even if they have limited time: the author's experience will guide them.

This book came into being through the clearsightedness of Jean-François Perrot, professor at the University of Paris VI and director of the CNRS Laboratory for Artificial Intelligence and Forms (LAFORIA). Having supervised Boizumault's thesis, Professor Perrot recognized in his work the interest and qualities required for a presentation that goes beyond the scope of specialists and addresses a large public. Developing the project took two full years.

It is understood that an experienced reader will probably not be reading these chapters. The presentation is, however, sufficiently well-organized to draw readers who possess some idea of logic programming.

Patrice Boizumault has above all achieved a scientific and industrial ethnological work: he has studied, analyzed, and dissected the Prolog implementations, from the first historical implementation written by Philippe Roussel in Algol-W to the Prolog-II implementation, via the Edinburgh Prolog; and, at the opposite end, the Australian and Japanese Prologs. He did not address Prolog-III, for the very reason that Alain Colmerauer reserves for himself its complete operational description, preserving the suspense until he completes this description.

The numerous "Prolog systems" that have been implemented result from the abundance of ideas with which their authors experimented, often successfully and efficiently, sometimes not. Most of the authors benefited from the experience acquired from the unspeakable number of Lisp dialects for which they harbor a touching nostalgia, associated with strictly Prologian finery. Patrice Boizumault has achieved here a synthesis that translates the know-how of the most talented programmers. He has succeeded in presenting the concepts, difficulties, and design limits, and in providing solutions and making them understandable to all, both the impetuous Prologian or the programmer concerned with understanding the technical foundations of his tool.

For the exposé to be complete, the proposals had to be demonstrated. This has been achieved—as for any software design—through the programming of the most delicate mechanisms. The reader can thus be convinced that the proposed solutions are operational and feasible.

The great merit of this book is that it allows the reader to discover that there is nothing mysterious about Prolog. The different design descriptions translate the personal visions of their authors—whether architectural or not—whereas this presentation is more impartial: it simplifies the problems and proposes and clarifies solutions.

However, the presentation goes beyond Prolog itself. The fundamental problems of its implementation—such as unification, control, backtracking, the *cut* rule (the famous *cut* or *goto* of the 1980s), and the compilation itself—are often ill-solved, if not merely patched up. This is especially true in knowledge-base systems. We can only advise the authors of expert systems tools to explore the solutions presented in this book.

Patrice Boizumault should be thanked for his competent and rigorous presentation of the major problems posed by the implementation of Prolog, for his elegant solutions, and especially for his answers to the programmer's inquiries.

From the abundant literature that deals with this topic, it was time for a conclusive presentation that addresses everyone to emerge. There is no doubt that this book will become an epoch-making reference in the Prolog saga, which is undoubtedly in its first episodes.

Marc Bergman

ACKNOWLEDGMENTS

I would especially like to thank Jean-François Perrot for his help, advice, and encouragement during the writing of this book. I would also like to thank Marc Bergman, Patrick Greussay, Yvon L'Hospitalier, and Eric Papon for their interest in this project. Finally, I thank Michel Billaud, Jean-Pierre Briot, Phillippe Codognet, Rémi Legrand, Guy Lapalme, Vincenzo Lola, Michel Quaggetto, François-Xavier Testard-Vaillant, and Jean Vaucher for their help in the preparation of this book.

Patrice Boizumault

THE IMPLEMENTATION OF PROLOG

INTRODUCTION

The concept of using predicate logic as a programming language appeared in the early 1970s as a result of the works of A. Colmerauer and R. Kowalski. The first Prolog interpreter (Prolog0) [CKPR72] was developed in 1972 at the university of Aix-Marseille by P. Roussel. This first implementation was written in Algol W and used a complete copying technique.

The power and simplicity of Prolog, together with a well-defined semantics based on predicate calculus, gave birth during the 1970s to numerous works and applications, thus promoting a wide variety of implementations. Among these, let us just recall:

- Marseille Prolog1, alias Prolog Fortran [BM73], [Rou75], which was the first implementation to use a term representation by structure sharing [BM72];

- Prolog-Dec 10—the first Prolog compiler—developed at Edinburgh University by D. H. Warren [War77]; and

- the works of M. Bruynooghe [Bru76], [Bru80] and C. S. Mellish [Mel80], who proposed an alternative terms representation: structure copying.

Clauses indexing first appeared in Prolog-Dec 10 [War77]. Last-call optimization was first introduced in 1977 by M. Bruynooghe [Bru80] in an implementation that used structure copying. It was then reconsidered by D. H. Warren in 1979 [War79], [War80] in the context of structure sharing. In 1983, Warren proposed a way to compile Prolog using a virtual machine, the WAM (Warren's Abstract Machine). The WAM has been abundantly referred to in other works and is the de facto standard for implementing Prolog compilers.

The early 1980s witnessed the appearance of a new generation of Prolog systems that offered the possibility of modifying the standard control strategy (depth first search from left to right of the *and/or* tree). These systems allow the user to specify explicitly the required control, whether through variable annotations or through a delayed evaluation.

Among all the works achieved in this field, let us mention:

- the *freeze*/2 predicate of Prolog-II [Col82b], [Kan82], [Can82];

- IC-Prolog and its variable annotations mechanism [CC79], [CCG82];

- the *wait declarations* of MU-Prolog [Nai85a], [Nai85b]; and

- the *when declarations* of NU-Prolog [Nai86].

Logic programming took a new turn in October 1981, after the Japanese scientific community announcement of the fifth-generation project, for which Prolog, or one of its derivatives, was to be the basic language for parallel processing computers.

Much research on parallelism has been undertaken since 1982. Parallel logic programming languages are usually classified into two main families [dKCRS89]. The first family is based on the guard concept and that of *don't-care nondeterminism*, which only retains the first clause whose guard is successfully executed. The synchronization of the *and* processes is performed by accessing shared logical variables. This is achieved either through variables annotation as in concurrent Prolog [Sha88], through *mode declarations* in Parlog [CC84] [CG85], or through input guards as in guarded Horn clauses [Ued85], [Ued86].

The second family is composed of nondeterministic logical languages that directly relate to Prolog for the property they share in providing multiple solutions. Parallelism can be implicit (*don't-know nondeterminism*) and its management transparent to the user [War87a], [War87b], or explicitly specified as in PEPSys [RS87].

Thus, in a first attempt, all the existing Prolog systems (and their derivatives) can be classified according to two main categories:

1. sequential Prologs, which can be further decomposed into:

   a. the *classical Prolog* systems, which use the standard control strategy

   b. the *Prolog+* systems, which allow modification of the standard control through variables annotation or through a delaying mechanism; and

2. the parallel Prologs.

Numerous works are currently under way to introduce the concept of constraint in the Prolog language [Col87], [DSH87], [Hen89], [JL87]. Under this approach, a Prolog program is still composed of facts and rules. However, in addition each clause contains a specific part that describes the set of constraints under which it must execute.

The purpose of this approach is to use constraints actively [Gal85] in order to reduce a priori the size of the search space, especially through the generation of new constraints from those already imposed.

Our study will be restrained to sequential Prologs without a specific constraints management mechanism.

A Great Variety of Sequential Prolog Systems

Even though the *Prolog+* systems are characterized by the new control capabilities they offer, *classical Prologs* do not lack diversity, whether apparent in their syntax, their built-in predicates, or their programming environment; or not so apparent in their implementation choices.

Indeed the Prolog systems developed throughout the world during the last fifteen years offer a wide syntax variety.

Some examples are:

- Clauses syntax: the "neck" symbol :- of Edinburgh Prologs [PPW78], [Per84b], the + and - symbols that precede literals in Prolog/P [Man84], the arrow -> of Prolog-II [Can82]; and

- Variables syntax: a variable name starts with an uppercase letter in Prolog-Dec10 and C-Prolog, is prefixed by a special character like "*" in Prolog/P, or has a particular syntax, as in Prolog-II.

Furthermore, these systems present a great variety of built-in predicates, interfacing capabilities, and programming environments.

This diversity is a result of the youth of the language. The standardization of the Prolog language is currently under way [DRM87]. The future standardized Prolog will strongly resemble Prolog-Dec10, which seems to be a de facto standard, at least in Anglo-Saxon countries.

All these systems also differ in a more subtle way—through their implementation. In this context, we can distinguish three main levels, by which the implementation of a *classical Prolog* can be characterized:

1. its working zone architecture,
2. its structured terms representation, and
3. its memory management efficiency.

The first and third levels are intimately related. Indeed, the working zone architecture of a *classical Prolog* system is dependent on the efficiency of the space-saving techniques that are to be implemented:

1. updating upon backtracking only,
2. updating upon return from deterministic calls, and
3. last call optimization.

As for the second level, it allows classification of the implementations according to the two terms representation modes: structure sharing and structure copying.

The Purpose of This Book

Taking into account the richness and diversity of the proposed implementations, this book is intended to sum up the implementation of a Prolog system, whether a *classical Prolog* or a *Prolog+*. Accordingly, we will extract the basic principles common to all the implementations and highlight the specific problems of the implementation of Prolog. We will then study the proposed solutions while comparing the different approaches. Finally, we will implement three different Prolog interpreters that will illustrate the problems and solutions we have described.

Global Approach

Our global approach is based on the observation that if a language is to be implemented, its basic mechanisms should first be well understood. The different implementation principles and techniques should then be studied before the implementation phase can be undertaken. We thus propose a general layout divided in three distinct parts:

- Part 1: Fundamental Principles of the Language
- Part 2: Principles and Techniques of Implementation
- Part 3: Implementations

We will first recall the fundamental principles of the language. Next, we will distinguish among three types of Prolog systems implementations, depending on the choices made at the levels of the terms representation, memory management, and possible control alternatives. In the last part, dedicated to implementations, we will implement three different interpreters that correspond to the three levels previously described.

All our examples and implementations will have the syntax and built-in predicates of Prolog-Dec10 and C-Prolog. However, our programs will consist only of Horn clauses—that is, we will not use the *or* operator (usually denoted by ;).

Two Main Progression Levels

Two main progression levels besides the global approach have been retained: the transition from *classical Prolog* to *Prolog+* and the transition from interpreted to compiled classical Prolog.

1. From *classical Prolog* to *Prolog+*:

Since *Prolog+* is an extension of *classical Prolog*, we will always consider the aspects relevant to *classical Prolog* before considering those relevant to *Prolog+*, whether these are principles, implementation techniques, or implementations:

a. fundamental principles: chapters 1 and 2 for *classical Prolog*; chapter 3 for *Prolog+*;

b. principles and techniques of implementation: chapters 4 to 9 for *classical Prolog*; chapter 10 for *Prolog+*;

c. implementations: chapters 11 and 12 for *classical Prolog*; chapter 13 for *Prolog+*;

2. *Classical Prolog*: from interpretation to compilation
Before dealing with Prolog compilation, it seems essential to understand first the choices made by the implementors in order to solve the specific problems of Prolog. Taking into account the diversity of the approaches, it seems important first

- to present the specific implementation problems,
- to describe the different solutions, and
- to compare these solutions.

During this first phase, we will voluntarily deal with interpretation. We will then naturally reach the compilation phase:

a. the first phase dealing with the presentation of the problems and the solutions (together with their comparison) will be the topic of chapters 4 through 8;

b. the second phase, relevant to compilation, will be covered in chapter 9.

Despite this classification, we will always keep the compilation goal in mind in chapters 4 through 8. Accordingly, we will make some choices that, beyond solving the problems in an interpretive context, will reveal their full potential and benefit in the context of compilation.

The Choice of an Implementation Language

All the implementations in this book will be written in Common-Lisp [Ste84].

First, it seems essential to use a single language in order to obtain a more unified presentation. Furthermore, this allows the direct reusability of the programs already developed in preceding chapters.

Moreover, Lisp is an excellent language for the implementation of Prolog, as the abundance of Prolog systems developed in Lisp reveals:

- LogLisp [RS82],
- Qlog [Kom82],
- LisLog [BDL84],
- LoVlisp [Gre83],
- LM-Prolog [KC84],
- Logis [Glo84].

All of these systems offer, at different levels, an integration between Lisp and Prolog. They permit the combination of the advantages of both languages under one programming environment, by allowing calls to Lisp from Prolog and calls to Prolog from Lisp. Those interested in finding more information on the languages that combine Lisp and Prolog can refer to [Din85] where they will find a complete survey of this topic.

Our approach will remain much more modest in this respect. Indeed, we will not try to design a new blend of Lisp and Prolog but rather will simply use Lisp as a Prolog implementation language.

Finally, the choice of Lisp as an implementation language assures continuity with our previous achievements [Boi85], especially the Vlog1 and Vlog2 systems initially written in Vlisp [Gre82], then ported [Loi87] to Common-Lisp [Ste84].

Detailed Plan

1. Part 1: Fundamental Principles of the Language
   We will first recall the principles of *classical Prolog*, separating unification (chapter 1) from control (chapter 2). We will then consider the possible enhancements in terms of control (chapter 3). Accordingly, we will consider three *Prolog+* proposals: Prolog-II [CKC83], MU-Prolog [Nai85a], and IC-Prolog [CC79],[CC80].

2. Part 2: Principles and Techniques of Implementation
   The transition from interpretation to compilation will be achieved in two steps:

   a. Step 1: Presentation of the problems, solutions, and comparisons.
      We will distinguish between two levels depending on the efficiency of the implemented memory managements.
      - Level 1: updating upon return from deterministic calls
        We will first present the control implementation together with the problems specific to backtracking (chapter 4). We will then consider the two terms representation modes: structure sharing

and structure copying (chapter 5). We will then reach the first level (chapter 6).

- Level 2: last-call optimization
  We will adapt this version to the requirements of the last-call optimization, thus building the second level (chapter 7). We will then conclude this first step (chapter 8) by describing the clauses indexing mechanism that permits natural reinforcement of determinism.

b. Step 2: Compilation

We will then have grasped all the elements required to consider the specific aspects of compilation (chapter 9). We will focus on the solution proposed by D. H. Warren in 1983, namely Warren's Abstract Machine [War83].

Among the *Prolog+* systems presented in chapter 3, we will only retain Prolog-II, thus setting our third and last level. Indeed, the *freeze* predicate is nowadays available on many Prolog systems such as SP-Prolog [SPP87] or Sicstus Prolog [Car87]. Furthermore, the implementation of such a delaying mechanism has been widely covered [Can84], [Boi86b], [Car87].

3. Part 3: Implementations

We will follow the evolution of Part 2 by respecting its three successive levels. We will thus implement three different interpreters: Mini-CProlog, Mini-WAM, and Mini-Prolog-II.

Mini-CProlog (chapter 10) considers the proposals of chapter 6. It implements the updating upon return from deterministic calls and uses a structure-sharing terms representation. Mini-CProlog is an interpreted version of Prolog-Dec10.

Mini-WAM (chapter 12) is the result of the considerations developed in chapter 7, and performs the last-call optimization while using structure copying. Mini-WAM can be considered an interpreted version of Warren's Abstract Machine.

Mini-Prolog-II (chapter 13) implements the delaying mechanism associated with the *dif*/2 and *freeze*/2 predicates of Prolog-II, by considering the principles stated in chapter 10. Mini-Prolog-II can be considered a Prolog-II implementation based on the architecture and memory management of Warren's Abstract Machine.

Chapter 14 will deal with the built-in predicates. We will especially focus on the implementation of *cut*.

Conclusion

Finally, we will conclude by considering the machines dedicated to the execution of sequential Prolog. Among the achievements made in this field since the 1980s, we will retain two proposals that differ in their goals and thus in their approach with respect to the implementation of Prolog:

1. the PSI machine (Personal Sequential Inference) developed in the context of the Japanese fifth-generation project [NYY*83], [FF86]; and

2. the MALI machine (Machine Adaptée aux Langages Indéterministes), developed by the IRISA of Rennes [BCRU84], [BCRU86].

We will then describe the implementation choices that were made for each of these two implementations and characterize them according to the problems and solutions described in Part 2.

Part I

Fundamental Principles of the Language

The purpose of this first part is to recall the fundamental principles of *classical Prolog* as well as those of *Prolog+* with delaying mechanism. As required, we will also introduce the problems that will be treated in the second part as well as some common terminology.

We will closely follow the second level of progression discussed in the general introduction, that is, the passage from *classical Prolog* to *Prolog+*. For this, we will first describe the fundamental principles of the Prolog language, by considering the notions of unification (chapter 1) and control (chapter 2). Then, we will discuss (chapter 3) the various propositions needed to improve the *classical Prolog* control.

Fundamental Principles of Prolog

We have deliberately separated the notions of unification and control. For each one, we will begin with the general concept and then show the restrictions or choices made by Prolog. The algorithms described are implemented in Lisp at the end of each chapter. For more details on the basic concepts, we will often refer to well-known books.

The first chapter is devoted to the discussion of unification. Our starting point will be the unification algorithm of J. A. Robinson [Rob65], [Rob71] and, as we shall see, the vast majority of Prolog systems choose not to make the occur check. We will then describe the problems that this poses and present the solution brought about by unification of infinite terms in Prolog-II [Col82a],[Col82b], [Kan82], [Can82]. We will close with the implementation of these three unification algorithms.

The subject of the second chapter is the *classical Prolog* control. We will assume familiarity with the principle of resolution as well as the fundamental results of predicate calculus. We will recall the notions of proof by refutation and of Horn clauses. We will then describe the resolution strategy used by Prolog in order to arrive at the notion of classical control with two selection rules: a rule for clause selection and a rule for literal selection. We will then define the notions of forward and backward moves, which will be represented by means of the traversal of an *and/or* tree. We will conclude with the implementation of a pure Prolog interpreter.

Improvements on the *Classical Prolog* Control

We will distinguish (chapter 3) four principal levels of improvement on the *classical Prolog* control [GL82]: the pragmatic control, the explicit control incorporated in the program, the explicit control separated from the program, and the automatic control. We will be particularly interested in the third level consisting of the explicit control incorporated in the program (*Prolog+*).

For this purpose, we will select three approaches:

- the *variables annotation* of IC-Prolog [CC79], [CC80], [CCG82];
- the *wait declarations* of MU-Prolog [Nai85a], [Nai85b];
- the predicate *freeze* of Prolog-II [Col82b], [Kan82], [Can82].

We will present each one separately and then we will compare the three approaches. Finally, we will conclude with some comments about the *when declarations* of NU-Prolog [Nai86], [Nai88].

CHAPTER 1

UNIFICATION

Unification is an essential part of any Prolog system. Unification is the process of determining whether two terms can be made identical by appropriate substitutions for their variables. The first unification algorithm was proposed by J. A. Robinson in 1965 [Rob65].

But a vast majority of Prolog systems do not use Robinson's unification algorithm. For reasons of efficiency, they deliberately choose not to perform the occur check (a variable can thus be unified with a term containing it).

This choice entails some problems from a theoretical point of view, since it destroys the soundness of the SLD-resolution strategy used by Prolog. From a practical point of view, a unification algorithm without the occur check may not terminate.

In order to solve these problems, Prolog-II [Col82b], [Can82], [Kan82] has proposed a new theoretical model taking into account the infinite terms created by the absence of occur check. Unifying two finite terms is replaced by solving a system of equations over rational trees.

After briefly recalling some definitions, we will describe the unification algorithm of J. A. Robinson. Then we will study the particular algorithm used by Prolog by indicating the problems that arise when the occur check is abandoned. We will then examine various proposals to remedy this situation, particularly Prolog-II's approach. Finally, we will close with the implementation of the various algorithms previously described.

1.1. Definitions

A Prolog term is a variable, a constant symbol, or a structured term. A structured term is either a list or a functional term. A functional term t is an expression of the form $f(t_1, ..., t_m)$ where f is a functional symbol, and t_i are terms. We will say that f is the functor of t whose arity is m. Lists can also be specified as functional terms using the functor '.' of arity 2. $[a, b, c]$ and $'.'(a,' '.'(b,' '.'(c, [])))$ denote the same list.

A substitution σ is a finite set of pairs $(variable, term)$ such that the same variable does not appear as the left side of two or more pairs.

Example:
$$\sigma = \{(Y, f(X, Z)); (X, a)\}.$$

We will say that the variable X (resp. Y) is bound to the term a (resp. $f(X, Z)$). On the other hand, the variable Z is said to be free, since it does not appear as the left side entry in any pair. Applying the substitution σ to the term t gives the term $t' = \sigma(t)$ obtained by replacing in t, all occurrences of variables by their related terms given by σ. t' is said to be an instance of t.

Example:
> let $t = g(X, Y)$ and σ the preceding substitution
> then $t' = \sigma(t) = g(a, f(a, Z))$.

A unifier of two terms t_1 and t_2 is a substitution σ such that $\sigma(t_1) = \sigma(t_2)$. If such a substitution exists, we say that the terms t_1 and t_2 are unifiable.

Examples:
> $t_1 = f(X, Y)$ and $t_2 = f(Y, g(a))$ are unifiable and the substitution
> $\sigma = \{(X, Y); (Y, g(a))\}$ is such that $\sigma(t_1) = \sigma(t_2)$.
> $t_1 = f(X, X)$ and $t_2 = f(a, b)$ are not unifiable.

A unifier σ of two terms t_1 and t_2 is said to be the most general unifier of t_1 and t_2 if and only if, for all unifiers θ of the two terms, there exists a substitution λ such that $\theta = \lambda \circ \sigma$.

Example:
> Let $t_1 = f(X, a)$ and $t_2 = f(Y, Z)$.
> Their most general unifier is $\sigma = \{(X, Y); (Z, a)\}$.
> If $\theta = \{(X, Y); (Y, a); (Z, a)\}$ then $\theta(t_1) = \theta(t_2)$
> but there exists $\lambda = \{(Y, a)\}$ such that $\theta = \lambda \circ \sigma$.

If two terms are unifiable then their most general unifier always exists. The first algorithm to calculate this was proposed by J. A. Robinson in 1965 [Rob65].

1.2. Robinson's Unification Algorithm

We will look at the unification algorithm of J. A. Robinson [Rob65], [Rob71] in its recursive form as described in [BC83]. Constants are functional terms of arity zero.

Given: two terms t_1 and t_2.
Result: a pair $(bool, \sigma)$ such that:
> -$bool = true$ if and only if the two terms are unifiable.
> -if $bool = true$ then σ is the most general unifier for t_1 and t_2.
Algorithm:
> if one of the two terms is a variable x, call the other term t.

```
         then if x = t
                 then bool := true; s := ∅
                 else if occur(x, t)
                         then bool := false
                         else bool := true; σ := {(x, t)}
                         endif
                 endif
         else    let t₁ = f(x₁, ..., xₘ) and t₂ = g(y₁, ..., yₙ)
                 if f ≠ g or n ≠ m
                         then bool := false
                         else i := 0
                             bool := true;
                             σ := ∅;
                             while i < n and bool do
                                 ι := i + 1;
                                 (bool, σ₁) := unify(σ(xₖ), σ(yₖ));
                                 if bool then σ := σ₁ ∘ σ endif
                             endwhile
                 endif
     endif
```

Thus, $unify(f(X, a, T), f(Y, Z, b))$ successively constructs the substitutions $\{(X, Y)\}$, then $\{(X, Y); (Z, a)\}$, and finally $\{(X, Y); (Z, a); (T, b)\}$, which is the most general unifier of the two terms whose most general common instance is $f(Y, a, b)$.

According to the algorithm of J. A. Robinson, a variable cannot unify with a term containing it. Thus, the two terms $f(X, Y)$ and $f(Y, g(X))$ are nonunifiable. This verification, called occur check, is carried out by the function $occur(x, t)$ which returns $true$ if and only if the variable x appears in the term t.

1.3. Prolog's Unification Algorithm

The vast majority of Prolog systems do not perform the occur check. Thus, contrary to Robinson's unification algorithm, a variable can be unified to a term containing it. After we describe the reasons for such a choice, we will examine the problems it entails.

1.3.1. Unification without Occur Check

The unification algorithm of Prolog does not perform the occur check. When unifying a variable x against a term te, the system does not check whether the variable x occurs in te. Thus, the two terms $f(X)$ and $f(g(X))$ become unifiable in Prolog with unifier $\{(X, g(X))\}$.

The main reason for this omission is a sensible gain in execution time. Indeed, the occur check is a costly operation since, before binding a variable x to a term te, one must scan the term te in order to determine whether x occurs in it. Without the occur check, unifying a variable x to a term te is done in constant time independent of the size of te. As noted by A. Colmerauer in [Col82a], the concatenation of two lists is quadratic in the length of the first list if a unification algorithm with an occur check is used, whereas it is linear without such a test.

So unifying a term against a variable, which can be considered the basic operation in Prolog, can take constant time only if the occur check is omitted. As quoted by Pereira in [Per84a], "The absence of occur check is not a bug or design oversight, but a conscious design decision to make Prolog into a practical programming language."

1.3.2. Resulting Problems

The absence of the occur check entails the unsoundness of the resolution strategy used by Prolog. Consider the following Prolog program:

```
unsound :- eq(Y,f(Y)).
eq(X,X).
```

The proof of unsound leads to a success unifying Y to f(Y) while unsound is not a logical consequence of the program.

Unification without occur check is likely to generate circular terms, which no longer belong to the universe of finite terms. So, such a unification algorithm may not terminate. In fact, any attempt to access an infinite term leads to a loop.

When trying to prove the goal eq(f(X,Y,X),f(g(X),g(Y),Y)), the unification algorithm will enter an infinite loop in attempting to unify X and Y in the substitution {(Y,g(Y)); (X,g(X))}.

This problem did not arise during the proof of unsound because the circular term created by unifying Y and f(Y) is never accessed. But attempting to write such a term will lead the Prolog printer to enter a loop:

```
| ?- eq(Y,f(Y)).
    Y=f(f(f(f(f(f(f(f(.......
```

These examples may appear artificial. In fact, experience shows that a vast majority of the problems usually solved in Prolog do not require the occur check. But there are useful programs in which the absence of occur check is a major drawback.

This problem appears most commonly in the use of difference lists [Llo86]. A difference list is a term of the form d(X,Y) that represents the difference between the two lists X and Y. d(X,X) denotes the empty list. The main interest of difference lists is that they can be concatenated in constant time.

Consider the following example from [SS87]. We want to implement queues so that insertion and deletion are constant time operations.

```
enqueue(Item, d(QS, [Item | QT]), d(QS, QT)).
dequeue(Item, d([Item | QS], QT), d(QS, QT)).
empty_queue(d(QS, QS)).
```

Every call to empty_queue with a nonempty list as argument proceeds erroneously by creating a circular term.

1.4. Solutions

A first solution is to bring in the unification algorithm with occur check, when necessary. The second approach consists in extending the theoretical Prolog model to take the infinite terms into account. This is the solution proposed by Prolog-II.

1.4.1. Bring in the Occur Check

A first remedy is to introduce into the Prolog system an option for unification with occur check. The user may then choose, depending on the circumstances, the suitable unification algorithm, that is, with or without occur check. Generally, it is this latter option that holds by default. This is the case for Prolog-P [BJM83], [Man84], with the options occur-on and occur-off and also for Prolog-Criss [Don83], [Lep85]. Nevertheless, in the default mode, the user still faces the same problems described above.

A more suitable approach is to have a preprocessor that is able to detect places in a Prolog program where infinite terms may be created. D. Plaisted [Pla84] has developed such a preprocessor, which adds checking code to these places in order to cause subgoals to fail if they create infinite terms.

1.4.2. Prolog-II's Solution

Prolog-II offers a completely different solution, which consists in taking advantage of, rather than rejecting, the infinite terms. As shown by A. Colmerauer [Col82a], [Col82b], infinite trees are natural representations for graphs, gram-

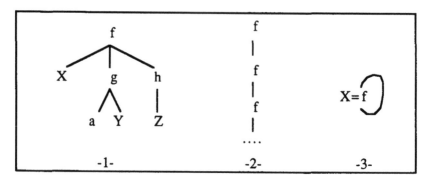

FIG. 1.1. Rational trees

mars, and flowcharts. Prolog-II extends the domain of the finite terms to rational trees (infinite terms) and accordingly replaces the notion of unification with that of the resolution of a system of equations over rational trees.

We will briefly present the basic notions of this approach with some examples. The interested reader can refer to [Col82b] and [CKC83], where a complete description of the Prolog-II theoretical model may be found.

1.4.2.1. Definitions

A tree is rational if and only if the set of its subtrees is finite. In particular, each finite tree is rational. Thus, $f(X, g(a, Y), h(Z))$ and $f(f(f...(X)...))$ are rational trees. In fact, the first one is a finite tree and the second one contains itself as its only subtree.

Since a rational tree has only finitely many subtrees, it can be represented simply by a finite diagram. All nodes having the same subtrees are merged together. So diagrams 2 and 3 of Figure 1.1 represent the same rational tree.

A tree assignment is a set X of the form $X = \{x_1 = t_1, ..., x_n = t_n, ...\}$ where the x_i's are distinct variables and the t_i's are rational trees. $te(X)$ denotes the tree obtained by replacing each occurrence of x_i in the term te by the corresponding tree t_i according to the tree assignment X.

The tree assignment X is said to be a solution of the system, possibly infinite, $\{p_1 = q_1, p_2 = q_2, ...\}$ if X is a subset of a tree assignment Y such that for all i, $p_i(Y) = q_i(Y)$.

A reduced system is a finite system of equations, without a cycle of variables, of the form $\{x_1 = t_1, ..., x_n = t_n\}$ where the x_i are distinct variables and the t_i are terms.

We define a cycle to be a nonempty system of the form $\{t_1 = t_2, ..., t_n = t_1\}$ where the t_i are distinct terms.

Finally, we define [Can84], [Can86] the representative of a term p in a system S without a cycle as follows: if there is an equation in S of the form $p = q$, then $rep(p, S) = rep(q, S)$ else $rep(p, S) = p$.

1.4.2.2. An Algorithm for Base Reduction

The algorithm for base reduction is used to transform a finite system of equations into an equivalent system, that is either reduced or trivially unsolvable. In fact, one can show [Col82b], [CKC83] that each system in reduced form is solvable (the solvability property) and that each solvable system is equivalent to a system in reduced form (the normal form property).

We will present an algorithm based on a version described by M. Van Caneghem in his thesis [Can84]: let T be the initial system. We try to transform the initial system T into an equivalent reduced system S.

$S := \emptyset$;
$E := \emptyset$; E saves the equations which are generated by the reduction
            namely all the pairs of subterms previously tried to unify
            E is called the stack of equations.
Reduce(T):
Given: a finite system of equations T.
Result: a pair (ok, S) such that
       -$ok = true$ if and only if T can be put in reduced form.
       -if $ok = true$ then S is a reduced system equivalent to T.
Algorithm:
if $T = \emptyset$ then $ok := true$
   else let $(s = t)$ be an equation of T
         let $s_1 = rep(s, S \cup E), t_1 = rep(t, S \cup E)$
         if $s_1 = t_1$
           then reduce($T - \{s = t\}$)
           else
              if one of the terms is a variable x, let y be the other term
                then $S := S \cup \{x = y\}$
                   reduce $(T - \{s = t\})$
              else
                let $s_1 = f(x_1, ..., x_n)$ and $t_1 = g(y_1, ..., y_m)$
                if $(f = g)$ and $(m = n)$
                  then $E := E \cup \{s_1 = t_1\}$
                     reduce $((T - \{s = t\}) \cup \{x_1 = y_1, ..., x_n = y_n\})$
                  else ok:=false
              endif

          endif
      endif
  endif

One of the important points in this algorithm is the calculation of the representative of a term by means of S and the stack of equations E. Indeed, this computation is done at each step of the reduction and is a function of the size of the stack of equations. We must therefore try to minimize the size of this stack. For a discussion of this problem and the comparison between different algorithms, we refer the reader to M. Van Caneghem's thesis [Can84].

Finally, such a unification algorithm requires a special printing program to output infinite trees [Piq84]. An interesting characteristic of this program is that it can eliminate common subtrees, thus producing a minimal infinite tree.

1.4.2.3. Example

Let us trace this algorithm with the example that we had chosen (see section 1.3.2) in order to bring forth the deficiencies of the unification algorithm without occur check.

Suppose we have to reduce the system $T = \{f(X, Y, X) = f(g(X), g(Y), Y)\}$
$S := \emptyset$;
$E := \emptyset$;
Step 1:
  Let $t = f(X, Y, X)$ and $s = f(g(X), g(Y), Y)$
  $t_1 = rep(t, S \cup E) = t$, $s_1 = rep(s, S \cup E) = s$
  Since t_1 and s_1 have the same functor and same arity,
  E becomes $\{f(X, Y, X) = f(g(X), g(Y), Y)\}$
  and T equals $\{X = g(X), Y = g(Y), X = Y\}$
  Step 2:
    Let $t = X$ and $s = g(X)$,
    $t_1 = rep(t, S \cup E) = X$, $s_1 = rep(s, S \cup E) = g(X)$
    Since t_1 is a variable, E does not change
    S becomes $\{X = g(X)\}$ and T equals $\{Y = g(Y), X = Y\}$
    Step 3:
      Let $t = Y$ and $s = g(Y)$,
      $t_1 = rep(t, S \cup E) = Y$, $s_1 = rep(s, S \cup E) = g(Y)$
      Since t_1 is a variable, E is unchanged
      S becomes $\{X = g(X), Y = g(Y)\}$ and T equals $\{X = Y\}$
      Step 4:
        Let $t = X$ and $s = Y$
        $t_1 = rep(t, S \cup E) = g(X)$, $s_1 = rep(s, S \cup E) = g(Y)$

Since the two functional terms have the same functor and arity, we add the equation to the stack.

E becomes $\{f(X, Y, X) = f(g(X), g(Y), Y), g(X) = g(Y)\}$

S is not changed and T equals $\{X = Y\}$

Step 5 :

Let $t = X$ and $s = Y$

$t_1 = rep(t, S \cup E) = g(Y)$ and $s_1 = rep(s, S \cup E) = g(Y)$

Since these two terms are identical, ok:=true.

At Step 5, using the stack of equations that saves all the pairs of subterms for which unifying was previously tried (in particular $g(X) = g(Y)$), the algorithm skips over the equation $X = Y$, avoiding entering a loop. The initial system is therefore solvable with associated reduced system $S = \{X = g(X); Y = g(Y)\}$.

1.5. Implementation

We will first implement the classical unification algorithm of Robinson and then deduce from it the one used by Prolog. Then we will give a version of the base reduction algorithm of Prolog-II.

For this, we will adopt the following representation of terms:

- a Prolog constant will be represented by its corresponding Lisp atom;

- a Prolog variable will be a Lisp atom whose name begins with the character '_'; and

- a Prolog functional term will be represented by a list of car, its functor, and of cdr the list of its subterms. Thus, $f(X, g(Y), Z)$ will be coded by (f _X (g _Y) _Z).

1.5.1. Robinson's Unification Algorithm

Let us implement in Lisp Robinson's unification algorithm (see section 1.2). The substitutions will be classically represented by A-lists, in the form of associations between variables and values, that is, $((var_1.val_1)...(var_n.val_n))$.

```
(defconstant marque #\_)
(defun var? (x)
    (and (symbolp x) (char= (char (string x) 0) marque)))
(defun value (x s) (cdr (assoc x s)))
(defun bound? (x s) (assoc x s))
(defun add (x val s) (acons x val s))
```

(appsub te s) applies the substitution s to the term te. We will distinguish three cases: te is a variable, te is a constant, and te is a functional term. For a bound variable, we must apply the substitution to its value. Indeed, the variable may be bound to a term containing variables. For a free variable or a constant, such a term is itself. A functional term is handled recursively:

```
(defun appsub (te s)
   (cond
      ((var? te)
         (if (bound? te s) (appsub (value te s) s) te))
      ((atom te) te)
      (t (cons (appsub (car te) s) (appsub (cdr te) s)))))
```

Let us use the occur check to determine if a variable x appears in a term te:

```
(defun occur? (x te)
   (if (atom te)
      (eq x te)
      (or (occur? x (car te)) (occur? x (cdr te)))))
```

Finally, let us construct the unification. The unify function calls unif with, initially, an empty substitution. unif returns either the most general unifier in the form of an A-list, or else the atom fail if the two terms x and y are not unifiable.

```
(defun unify (x y) (unif x y ()))
(defun unif (x y s)
   (cond
      ((eql x y) s)
      ((var? x) (if (occur? x y) 'fail (add x y s)))
      ((var? y) (if (occur? y x) 'fail (add y x s)))
      ((or (atom x) (atom y)) 'fail)
      (t (let ((news (unif (car x) (car y) s)))
         (if (eq news 'fail)
            news
            (unif (appsub (cdr x) news)
                  (appsub (cdr y) news) news))))))
```

For a variable, we perform the occur check and depending on the result, we produce a failure or generate a new substitution. Two stuctured terms are unified recursively by propagating the effects of the generated substitutions. Two constants or two identical variables unify directly whereas two distinct constants give rise to a failure.

Let us close with the test[1] of the algorithm on some examples.

```
? (unify '(f _X _Y)  '(f a b) )
= ( (_Y . b) (_X . a) )
? (unify '(f _X _X)  '(f a b) )
= fail
? (unify '_X '(f _X) )
= fail
```

---

[1] ? is the Lisp prompt and = indicates the result of the evaluation.

```
? (unify '(f (g _X _Y _U) _U)  '(f (g a _U _Z) _Y) )
 = ( (_U . _Z)  (_Y . _U) (_X . a) )
```

The first two examples are very simple. The third one fails because of the occur check. The fourth one shows why it is necessary to apply the substitution recursively on the value of a bound variable. In this case, _Y is bound to _U, itself bound to _Z.

1.5.2. Prolog's Unification Algorithm

1.5.2.1. Version 1

One solution would be to take up the previous version, dropping the occur check. Only the definition of unif is changed.

```
(defun unif (x y s)
   (cond
      ((eql x y) s)
      ((var? x) (add x y s))
      ((var? y) (add y x s))
      ((or (atom x) (atom y)) 'fail)
      (t (let ((news (unif (car x) (car y) s)))
         (if (eq news 'fail)
            news
            (unif (appsub (cdr x) news)
                  (appsub (cdr y) news) news))))))
```

Now, (unify '_X '(f _X) ()) succeeds by generating the substitution ((_X.(f _X))).

But this solution has the effect of applying the substitutions beforehand to those parts of the terms not yet considered, as is shown by the two calls made to appsub in the definition of unif. This systematic application of the substitution is relatively costly because it requires the construction of two new terms to be substituted at each recursive call to unif.

1.5.2.2. Version 2

Let us now look at a second version that does not use systematic substitution on the remaining parts of the terms. The basic primitives remain unchanged:

```
(defconstant marque #\_)
(defun var? (x)
   (and (symbolp x) (char= (char (string x) 0) marque)))
(defun value (x s) (cdr (assoc x s)))
(defun bound? (x s) (assoc x s))
(defun add (x val s) (acons x val s))
```

Because substitutions are not applied systematically, an occurrence of a
bound variable no longer vanishes from the rest of the term. Let us study[2]
this new behavior of the function unif with an example.

```
(unify '(f _X _Y _Z _X) '(f _Y _Z _T a))
  ----> unif t1=(f _X _Y _Z _X) t2=(f _Y _Z _T a) s=()
  ----> unif t1=f   t2=f s=()
  ----> unif t1=(_X _Y _Z _X) t2=(_Y _Z _T a) s=()
  ----> unif t1=_X t2=_Y s=()
  ----> unif t1=(_Y _Z _X) t2=(_Z _T a) s=((_X . _Y))
  ----> unif t1=_Y t2=_Z s=((_X . _Y))
  ----> unif t1=(_Z _X) t2=(_T a) s=((_Y . _Z) (_X . _Y))
  ----> unif t1=_Z t2=_T s=((_Y . _Z) (_X . _Y))
  ----> unif t1=(_X) t2=(a) s=((_Z . _T) (_Y . _Z) (_X . _Y))
  ----> unif t1=_X t2=a s=((_Z . _T) (_Y . _Z) (_X . _Y))
  ----> unif t1=() t2=() s=((_T . a) (_Z . _T) (_Y . _Z) (_X . _Y))
  unif <---- ((_T . a) (_Z . _T) (_Y . _Z) (_X . _Y))
= ((_T . a) (_Z . _T) (_Y . _Z) (_X . _Y))
```

Since the second occurrence of the variable _X no longer disappears as a
result of the substitution {_X=_Y} during the unification between _X and a, one
must trace the list of bindings _X = _Y = _Z = _T to obtain its value.

We will call a chain of bindings of length n any sequence of bindings
$(var_1, var_2), (var_2, var_3), ..., (var_n, var_{n+1})$ where var_{n+1} is either free or
bound to a nonvariable term. In the previous example, the chain of bindings on
_X is of length 3.

We define the function ult, which runs through the binding chains in order
to determine their end point, by extending with the use of the primitive val the
notion of value to an arbitrary term:

```
(defun val (te s)
   (if (var? te) (ult te s) te))
(defun ult (v s)
   (if (bound? v s) (val (value v s) s) v))
```

The procedure described here by the function ult, which consists in running
through a binding chain in order to determine its endpoint, is usually known as
dereferencing.

```
(defun unify (x y)
   (unif x y ()))
(defun unif (x y s)
   (let ((x1 (val x s)) (y1 (val y s)))
      (cond
         ((eql x1 y1) s)
```

---

[2] ----> denotes a call, while <---- indicates a return. Since unif is tail-recursive, we just
indicate the last return.

```
((var? x1) (add x1 y1 s))
((var? y1) (add y1 x1 s))
((or (atom x1) (atom y1)) 'fail)
(t (let ((news (unif (car x1) (car y1) s)))
    (if (eq news 'fail)
        news
        (unif (cdr x1) (cdr y1) news)))))))
```

The definition of the function unif is the same as the previous one, but this time we do not access the values of the variables unless necessary. Moreover, direct calls to the function val simplify its writing and avoid recursive calls. Let us compare the behavior of these two versions on the same example:

```
? (unify_v1 '(f _X _Y _Z) '(f (g _X1 _X1) (h _X _X) (i _Y _Y)))
= ( (_Z . (i (h (g _X1 _X1) (g _X1 _X1)) (h (g _X1 _X1)
    (g _X1 _X1))))
  (Y .  (h (g _X1 _X1) (g _X1 _X1)))
  (_X . (g _X1 _X1)) )
? (unify_v2 '(f _X _Y _Z) '(f (g _X1 _X1) (h _X _X) (i _Y _Y)))
= ( (_Z . (i _Y _Y)) (_Y . (h _X _X)) (_X . (g _X1 _X1)) )
```

In the first case, the substitutions are applied to the remaining terms. In the second, we merely perform the variable-value bindings without propagating the effect of the substitutions unless necessary.

1.5.3. Prolog-II's Reduction Algorithm

Let us implement in Lisp the reduction algorithm of Prolog-II described in section 1.4.2. For this, we will continue to use the representation of terms as in the previous two sections.

The initial system sys, the stack of equations e, as well as the resulting reduced system s will be coded using A-lists. Thus, sys, e, and s will be of the form ((g1 . d1) ... (gn . dn)), where s has variables only on the left side.

Let us first perform the calculations for the representative of a term (see section 1.4.2):

```
(defun rep (x s e)
   (if (var? x) (right (right x s) e) (right x e)))

(defun right (g sys)
   (if (bound? g sys) (right (value g sys) sys) g))
```

To determine the representative of a term x, we go over the reduced subsystem s as well as the stack of equations e. If x is a variable we look for a first representative t1 in s, and then for a representative of t1 in the stack of equations e. For a nonvariable term, we restrict the search to e alone since all equations in s have a variable on the left side.

The primitive `right` runs over a system of equations `sys` looking recursively for the right end associated with a given left-hand member g. Its behavior is very close to that of the function `ult` (see section 1.5.2).

```
(defun reduce (sys e s)
   (if (null sys)
       s
       (let ((t1 (rep (caar sys) s e))
             (s1 (rep (cdar sys) s e)))
          (cond
            ((equal t1 s1) (reduce (cdr sys) e s))
            ((var? t1) (reduce (cdr sys) e (add t1 s1 s)))
            ((var? s1) (reduce (cdr sys) e (add s1 t1 s)))
            ((or (atom s1) (atom t1)) 'fail)
            (t (if (eq (car t1) (car s1))
                   (let ((a (reverse (split (cdr t1) (cdr s1) nil))))
                      (if (null a)
                          'fail
                          (reduce (append a (cdr sys))
                                  (add t1 s1 e)
                                  s)))
                   'fail)))))))
```

`reduce` implements the reduction of a system `sys` with the help of the stack of equations e and the reduced subsystem s (see section 1.4.2). When the system `sys` is empty, `reduce` succeeds and returns the reduced system s. If it is nonempty, we choose the first equation in `sys` and search the representatives of its left and right hand sides. If they are equal, we continue with the remaining system. If not, and if one of them is a variable, we continue with the remaining system after adding the new equation to the reduced system. Two different constants lead to a failure.

We still have to treat the case of two functional terms. If the functors are identical, we construct the new system associated with the two lists of arguments by verifying the arities (function `split`). We then continue the reduction with a new system consisting of the remainder of the old system to which the splitting of the two lists of arguments has been added, with a new stack of equations consisting of the old plus the equation just treated, and with the current reduced system (see section 1.4.2).

```
(defun split (x y r)
   (cond
      ((and (null x) (null y)) r)
      ((or (null x) (null y)) nil)
      (t (split (cdr x) (cdr y) (add (car x) (car y) r))))))
```

The function `split` carries out the transformation of an equation between two functional terms, at the same time checking that they have the same number of arguments. If the arities are different, `split` returns the empty list by default.

Let us see the tracing of `reduce` on the example of section 1.4.2.

```
reduce ---> sys = ((f _X _Y _X) . (f (g _X) (g _Y) _Y))
    e=()
    s=()
reduce ---> sys = ((_X . (g _X)) (_Y . (g _Y)) (_X . _Y))
    e= (((f _X _Y _X) . (f (g _X) (g _Y) _Y)))
    s= ()
reduce ---> sys = ((_Y . (g _Y)) (_X . _Y))
    e = (((f _X _Y _X) . (f (g _X) (g _Y) _Y)))
    s = ((_X . (g _X)))
reduce ---> sys = ((_X . _Y))
    e= (((f _X _Y _X) . (f (g _X) (g _Y) _Y)))
    s = ((_Y . (g _Y)) (_X . (g _X)))
reduce ---> sys = ((_X . _Y))
    e= (((g _X) . (g _Y))((f _X _Y _X) . (f (g _X) (g _Y) _Y)))
    s = ((_Y . (g _Y)) (_X . (g _X)))
reduce ---> sys = ()
    e= (((g _X) . (g _Y))((f _X _Y _X) . (f (g _X) (g _Y) _Y)))
    s = ((_Y . (g _Y)) (_X . (g _X)))
reduce <----  ((_Y . (g _Y)) (_X . (g _X)))
```

There remains to define the unification of two terms x and y by reduction of the system consisting of the equation x=y:

```
(defun unify (x y)
    (reduce (add x y ()) () ()))
```

CHAPTER 2

RESOLUTION AND PROLOG CONTROL

Every Prolog system can be considered primarily as a theorem-proving system [Rob65] whose common application consists of asking queries in the context of a universe defined in clausal form.

The answer to a query is obtained by refutation, that is, by showing that the system consisting of the initial clauses, to which the negation of the query is added, is unsatisfiable. This refutation uses a particular strategy of resolution [Rob65] called SLD-resolution [AE82].

The Prolog control implements such a strategy of resolution by choosing a particular computation rule (always selecting the left-most literal in the current resolvent) and a particular search rule (trying the clauses in their order of appearance). The *classical Prolog* control, defined in this way, can then be represented simply by a depth-first search from left to right on an *and/or* tree.

First, we will recall briefly how Prolog implements a refutation on a set of Horn clauses. Next, we will describe the particular strategy of resolution used, by specifying the choices made for the computation rule and for the search rule. We will then characterize all proofs in terms of forward and backward moves. Third, we will represent the Prolog control in the form of a particular traversal of an *and/or* tree, recalling the notions of a search tree and a proof tree. Finally, we will conclude with the implementation of a pure Prolog microinterpreter.

2.1. Refutation on Horn Clauses

We will not discuss here the principle of resolution [Rob65] or the predicate calculus of first order logic amply described in [CL73], [Lov78], and [Del86]; rather we will recall how Prolog performs a refutation on a set of Horn clauses.

2.1.1. Horn Clauses

An atomic formula (or literal) is an expression of the form $p(t_1, ..., t_n)$ where p is a predicate symbol of arity n, and t_i $i\epsilon[1..n]$ are terms.

A clause C of predicate calculus is a universally quantified expression of the form[1] $B_1 \lor ... \lor B_m \leftarrow A_1 \land ... \land A_n$ $(0 \leq m, n)$ or, equivalently, in normal disjunctive form: $B_1 \lor ... \lor B_m \lor (\sim A_1) \lor ... \lor (\sim A_n)$ where the B_i, $i\epsilon[1..m]$

---

[1]The logical connectives $\lor$, $\land$, $\sim$, and $\leftarrow$ have their usual meanings.

and the A_i, $i\epsilon[1..n]$ are atomic formulas. The B_i are said to be positive literals, and the A_i negative ones.

A clause C is a Horn clause if and only if $m \leq 1$, that is, C contains at most one positive literal.

There are four types [Kow74], [Kow79] of Horn clauses depending on the values of m and n:

- Rule: $B \leftarrow A_1 \wedge ... \wedge A_n$ ($m = 1$ and $n > 0$)
  $\forall x_1, ..., x_k$ variables appearing at least once in $A_1, ..., A_n, B$, if $A_1 \wedge ... \wedge A_n$ is true then B is true.
- Fact: $B \leftarrow$ ($m = 1$ and $n = 0$)
  $\forall x_1, ..., x_k$ variables appearing at least once in B, B is true.
- Denial: $\leftarrow A_1 \wedge ... \wedge A_n$ ($m = 0$ and $n > 0$)
  $\forall x_1, ..., x_k$ variables appearing at least once in $A_1, ..., A_n$, $A_1 \wedge ... \wedge A_n$ is false.
- Empty Clause: $\square$ ($m = 0$ and $n = 0$)
  always interpreted as false.

A definite clause is a clause that has exactly one positive literal.

Rules and facts are definite clauses.

By restricting to Horn clauses, we guarantee that there are no disjunctions in the conclusion section. Thus, the clause $(p \vee r) \leftarrow q$ has no direct equivalent in the form of Horn clauses; on the other hand, the formula $p \leftarrow (q \vee r)$ is equivalent to the conjunction of the following two clauses: $\{p \leftarrow q; p \leftarrow r\}$.

Nevertheless, Horn clauses are not restrictive. Their computational strength is the same as that of the classical models. For a complete survey of this subject, we refer the reader to chapter 8 of the book by C. J. Hogger [Hog84].

2.1.2. The Knowledge Base

The knowledge base of a Prolog system is a conjunction of definite clauses. Since each clause C contains one, and only one, positive literal H, we can easily regroup them in subsets according to the predicate symbol of H. This entails a simple organization of the knowledge base.

Consider, as an example, the beginning of a description of a hand of thirteen cards in bridge:

```
nb_of_cards(spades,5).
nb_of_cards(hearts,5).
nb_of_cards(diamonds,3).
nb_of_cards(clubs,0).
void_in(Coul) :-
    nb_of_cards(Coul,0).
```

```
singleton(Coul) :-
    nb_of_cards(Coul,1).
```

The predicate nb_of_cards/2 is defined by four facts, while void_in/1 and singleton/1 require only one rule each.

Each rule is divided into two parts: the head consisting of the positive literal and the body consisting of the rest of the clause formed by negative literals.

2.1.3. A Proof by Refutation

A common application of Prolog consists of asking queries about the universe defined by the knowledge base.

A query of the form ?- $Q_1, ..., Q_n$. in which the variables $x_1, ..., x_k$ appear, means: are there values for $x_1, ..., x_k$ such that $Q_1 \wedge ... \wedge Q_n$ is a logical consequence of the knowledge base.

The query ?- void_in(X). asks for values of X for which this statement is true taking into account the knowledge base:

```
nb_of_cards(spades,5).
nb_of_cards(hearts,5).
nb_of_cards(diamonds,3).
nb_of_cards(clubs,0).
void_in(Coul) :-
    nb_of_cards(Coul,0).
singleton(Coul) :-
    nb_of_cards(Coul,1).
```

To answer such a query, Prolog will attempt a refutation, that is, show that the system consisting of the knowledge base to which the negation of the query has been added (Horn clause of type denial), is unsatisfiable (nonmodelizable [Her30]).

The answer to the query ?- void_in(X). is obtained by proving that the following system is unsatisfiable:

```
not void_in(X)
void_in(Coul)   or   not nb_of_cards(Coul,0)
singleton(Coul)  or   not nb_of_cards(Coul,1)
nb_of_cards(spades,5)
nb_of_cards(hearts,5)
nb_of_cards(diamonds,3)
nb_of_cards(clubs,0)
```

This proof uses the principle of resolution [Rob65]. The implementation of a particular resolution method leads, by the process of unification, to the binding of the variables appearing in the query to the terms of the universe defined by the knowledge base. Thus, for the previous example, the variable X will be bound to the atom clubs. Moreover, this is the only binding for which the system is unsatisfiable.

Therefore, a first approach to Prolog consists of expressing knowledge in clausal form and then asking queries about the universe defined in this way. Prolog will then either produce a yes or no answer or exhibit values of the variables appearing in the query for which the query is a logical consequence of the knowledge base.

2.2. A Particular Resolution Strategy

Prolog uses a particular resolution strategy called SLD-resolution [AE82], [Llo86]. We will first introduce the SLD-resolution and related fundamental results, then we will describe the Prolog control defined by the choices of a computation rule and a search rule.

2.2.1. An SLD-Resolution

SLD-resolution represents linear resolution with selector function on definite clauses. A SLD-resolution is a strategy of linear and input resolution [CL73], [GN87], which uses a fixed rule for literal selection.

2.2.1.1. Definition

An SLD-resolution strategy consists of starting, first, with the denial C_0 consisting of the negation of the query, and then deriving at each step i $(i > 0)$ a new resolvent C_i

- obtained by the resolution of
    - C_{i-1} (linear deduction)
    - a clause C of the knowledge base (input deduction);
- using a fixed selection rule that determines uniquely the literal of C_{i-1} that will be resolved.

Consider the beginning description of a bridge hand, where only spades and hearts are included.

```
card(ace,spades).
card(queen,spades).
card(jack,spades).
card(3,spades).
card(king,hearts).
nb_of_cards(spades,4).
nb_of_cards(hearts,1).
honour(ace).
```

```
honour(king).
honour(queen).
honour(jack).
honour(10).
singleton(Coul) :-
    nb_of_cards(Coul,1).
honour_sing(H,C) :-
    card(H,C),
    honour(H),
    singleton(C).
```

Figure 2.1 shows a derivation of the empty clause proving that the king of hearts is a singleton honor. We start with the denial consisting of the negation of the query, and then at each step, a resolvent C_i is derived by resolution of C_{i-1} and a clause C of the knowledge base. We have systematically chosen the left-most literal of each resolvent.

Finally, we close with a remark about the input nature of the resolution. The restriction to Horn clauses, together with the choice of a linear resolution (with, at its head, the negation of the query), leads to the derivation of resolvents where all the literals are negative. The second clause must be chosen in the knowledge base that contains the only clauses having positive literals.

2.2.1.2. Literal and Clause Selection

The actual implementation of the resolution method described above is carried out by determining a rule for literal selection and a rule for clause selection.

Indeed, at each step i, we have two degrees of freedom:

- the choice of the literal L_j within the current resolvent R_{i-1} to which the resolution is applied, or

- the choice of a clause C of the knowledge base (whose head unifies with L_j) which will produce the new resolvent R_i.

We know [AE82], [Llo86], that whatever rule is chosen to select the literal, the completeness of the SLD-resolution is preserved. Moreover, the choice of the clause is also immaterial, if one can guarantee that all clauses that are eligible for a unification have been considered, no matter what the order.

2.2.2. Prolog Control

The classical Prolog systems always choose the left-most literal in the current resolvent and always consider the clauses according to their order of appearance in the knowledge base.

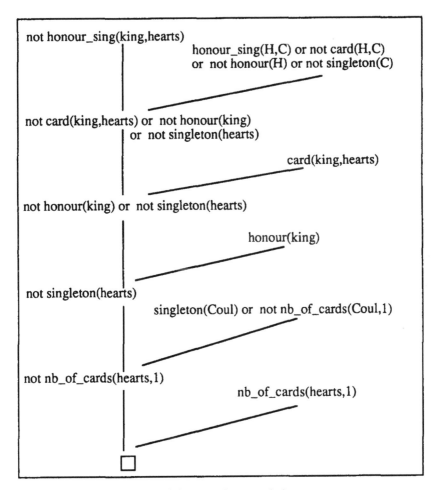

FIG. 2.1. Linear input resolution

2.2.2.1. Computation Rule

Prolog always selects the left-most literal of the current resolvent. Let $R_{i-1} = B_1...B_n$ be the current resolvent. The literal selection rule will choose B_1. Let $C = L_1...L_n$ be a clause of the base such that B_1 and L_1 unify to produce the most general unifier θ. Then the new resolvent R_i will be $(L_2...L_m B_2...B_n)\theta$.

This is the strategy used in the proof of Figure 2.1, where resolution is always applied to the left-most literal.

The process that performs the transition from one state of the derivation to the next will be called forward. By associating time with the proof, the forward process marks the passing of time from instant t to instant $t + 1$ by derivation of the resolvent R_{t+1} from R_t.

2.2.2.2. Search Rule

Prolog always selects the clauses that are eligible for the unification according to their order of appearance in the knowledge base, that is, in their sequential order of definition.

This choice will of course affect the order in which the answers to a given query are supplied. The following example illustrates this:

```
| ?-  listing(card).
   card(ace,spades).
   card(queen,spades).
   card(jack,spades).
   card(3,spades).
   card(king,hearts).
| ?-  card(V,C).
   V=ace      C=spades
   V=queen    C=spades
 • V=jack     C=spades
   V=3        C=spades
   V=king     C=hearts
```

We will say that there is a choice point associated with the goal B, if, when a clause is selected by the forward process to prove B and it satisfies the unification, the remaining set of clauses is nonempty.

When a failure occurs during the proof of a goal B (no more clause whose head unifies with B), it is then necessary to backtrack in order to reconsider the last choice that was made. This is done by consulting the latest choice point. A new proof of the goal that generated it is then attempted with the remaining set of clauses by restarting the forward process.

Thus backtracking is always done against time. In chronological terms, we pass from instant t_i to an earlier instant t_j, whereas the forward process marks the passage of time from t_i to t_{i+1}. These two movements forward and backward in time have been symbolized by Alain Colmerauer's clock [Col82b], [Col84].

2.3. Tree Representation

The Prolog control proceeds by successive trials in order to determine the set of all solutions to a given problem. This process can be represented by a depth-first search from left to right of an *and/or* tree.

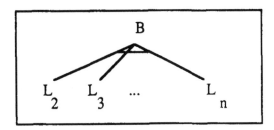

FIG. 2.2. *And* node

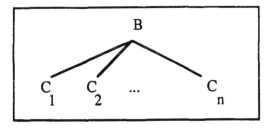

FIG. 2.3. *Or* node

2.3.1. Representation

The *and/or* trees used in artificial intelligence permit the representation of an important category of algorithms known as successive trial algorithms [Win77], [Nei79].

2.3.1.1. The *And* Nodes

And nodes are associated with the following situation: to solve a problem P, we are led to solve the more elementary subproblems P_1 and P_2 and ... P_n. For Prolog, this means: to prove a goal B using a clause $C = L_1 L_2 ... L_n$ whose head literal L_1 unifies with B, one must prove successively all the literals $L_i (i > 1)$ appearing in the body of clause C. *And* nodes will be denoted as described in Figure 2.2.

2.3.1.2. The *Or* Nodes

Or nodes are associated with the following situation: to solve a problem P, it is sufficient to solve one of the equivalent problems P_1 or P_2 or ... P_n. For Prolog, this means: to prove a goal B whose associated set of clauses is $P_q = C_1 ... C_n$, it is sufficient to prove the body of a clause $C_i, (1 \leq i \leq n)$, whose head unifies with B. *Or* nodes will be denoted as described in Figure 2.3.

2.3.1.3. Search Tree

The search tree describes the set of all possible proofs for a goal [AE82], [Emd84].

We will simply represent it by an *and/or* tree in the following way:

- To each rule, we associate an *and* node containing its head, and whose children will be the literals appearing in its body (conjunction).
- The *or* nodes will be associated with choosing a clause to solve a goal.

Figure 2.3.1.3 shows the proof of the query ?- honour_sing(king, hearts). There are three *or* nodes arising from the definition of the predicates card/2 and honour/1 (five facts each) and the predicate nb_of_cards/2 (two facts). All the other predicates are defined by one and only one rule. Each leaf of the tree corresponds to an elementary problem, that is, to a possibility of resolution by a fact. A success is symbolized by □, and a failure by △.

2.3.2. Depth-First Traversal from Left to Right of an *And/Or* Tree

The Prolog control can be represented by a depth-first traversal from left to right of an *and/or* tree. We will first describe such a traversal and then we will define the notion of a proof tree.

2.3.2.1. The Traversal

The two rules of literal and clause selection define completely the traversal of the search tree. Indeed, since the clauses are considered in the order of their

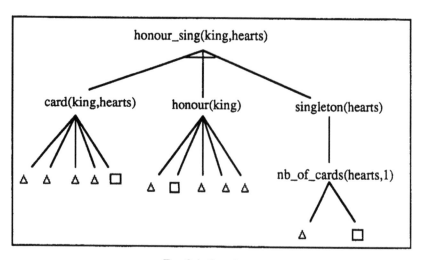

FIG. 2.4. Search tree

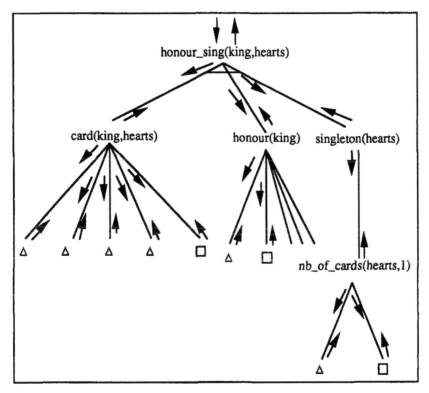

FIG. 2.5. Tree traversal

appearance, the traversal of the *or* branches (*or* control) is carried out depth-first, with a systematic backtracking to the latest choice point. Moreover, taking into account the rule for literal selection, the *and* branches (*and* control) are considered sequentially in a traversal from left to right.

Let us look at an example (Figure 2.5) by following the traversal used by Prolog on the search tree of Figure 2.4.

The traversal shown corresponds to the search by Prolog of the first solution (which, moreover, will be unique since the only *or* node, associated with honour(king), not entirely investigated, cannot produce other successes).

Let us do a concise tracing (Figure 2.6) of the traversal in indented form, by showing the proofs performed and the successes obtained (the various attempts in the *or* branches are not indicated). The traversal of the *and/or* tree is therefore done in depth-first order from left to right.

Here we realize the difficulty in accurately accounting for the Prolog control, even on a simple example. Indeed, the stream of information is very important because of the backtracking. It is therefore necessary to have good tracing tools

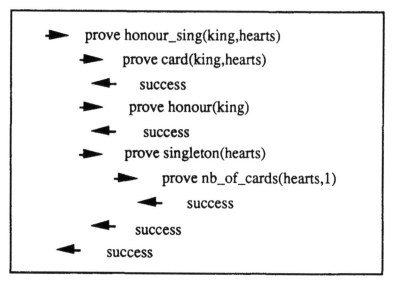

FIG. 2.6. Tracing

for recovering parts of proofs so as to allow the location, analysis, and correction of program errors. To learn more about the problems of tracing in Prolog, we refer the reader to the following articles: [Byr79], [Byr80], [Eis84], [Boi84], [FGP*85], [Con86], and [Duc86].

2.3.2.2. Proof Tree

The proof tree describes the proof of a goal without taking into account the backtrackings (only the successes produced in the proof of the goal are mentioned, not the failures that eventually occur). We will of course represent it in the form of a tree.

Figure 2.7 shows the proof tree associated with the search tree of Figure 2.5. The summary trace of Figure 2.6 actually describes the traversal of this proof tree, since we have not included the attempts at a proof leading to a failure.

Up until now we have been speaking in terms of the traversal of an *and/or* tree, assuming that it had already been constructed. In fact, Prolog will construct this tree step by step as the proof proceeds.

2.3.3. Completeness?

The Prolog control strategy derived from the two rules for literal and clause selection no longer preserves the completeness of the SLD-resolution.

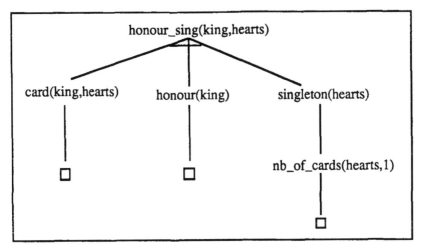

FIG. 2.7. Proof tree

2.3.3.1. The Problem

Prolog depth-first traversal will in some cases enter an infinite loop by systematically reusing the same clause without ever considering the others.

```
integer(s(N)) :-
    integer(N).
integer(zero).
```

Consider the classical definition of the integers in Prolog and ask the query
|?-integer(X). Prolog will then enter an infinite loop without giving any
answer. The first clause of the knowledge base is always chosen, whereas the
second one is never chosen, as shown by Figure 2.8.

However, there exists a derivation leading to the empty clause. To see this,
it is enough to consider directly the second clause, as shown by Figure 2.9.

Nevertheless, completeness is preserved when the search tree is finite
[Del86]. In this case, the Prolog system is sure to find all the derivations leading
to the empty clause.

To learn more about the detection problem of infinite loops, we refer the
reader to [Din79], [Din80], and [Bes85], as well as to the solution proposed
by Loglisp [RS80b], [RS80a], [RS82], which consists in internally limiting the
depth of the proof tree.

2.3.3.2. Remarks

Not only is writing Prolog programs declarative, as shown in the previous small
example; we must take into account the control for defining the set of clauses

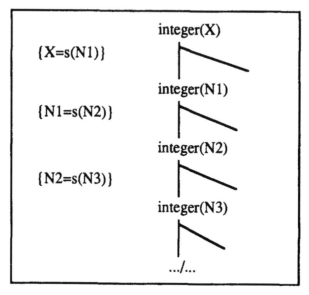

FIG. 2.8. Leading to an infinite loop

associated with a predicate.

In the definition of integer/1, any Prolog programmer would have reversed the order of the clauses. In this form, the Prolog system will then be able to "provide" the infinity of answers to the query, that is, zero and all its successors.

Thus, we see that writing Prolog programs requires a good understanding of control. Moreover, systematically exploring all the *or* branches can lead to a combinatorial explosion. It then becomes necessary to use primitives that

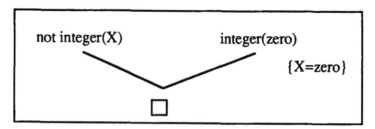

FIG. 2.9. Successful derivation

control the traversal of these branches, the most famous of which is the *cut*. Chapter 14 will be especially devoted to built-in predicates.

2.4. Implementation of a Pure Prolog Interpreter

Let us implement in Lisp a pure Prolog microinterpreter. We will first describe the syntax and the organization of the knowledge base. Then we will solve the problem arising from the multiple use of the same clause. We will then implement the Prolog control. Next, we will introduce the exhaustive search for solutions as well as the interpreter loop. Finally, we will close with a brief review of unification.

2.4.1. Syntax and Organization of the Knowledge Base

To simplify the reading and coding of Prolog clauses by the interpreter, we will use a syntax "à la Lisp." Rules and facts will be represented by lists of literals whose car is the positive literal. A query will be a list of literals. A literal is a list of car, the predicate symbol, and of cdr, the list of its arguments.

We will use the same representation of terms used in section 1.5.

- Constants will be Lisp atoms;
- Variables will be Lisp atoms whose name begins with the character '_';
- Functional terms will be Lisp lists of car, the functional symbol, and of cdr, the list of arguments;
- To represent Prolog lists, we will use the notation from Lisp, including dotted list with the macro character '.'; so [a,b,c] and [X | Y] will be coded as (a b c) and (_X . _Y).

We define some classical primitives for the manipulation of lists, such as membership of an element in a list, concatenation of two lists, naive reverse, deletion of an element and the resulting definition of list permutations.

```
((member _X (_X . _Y)))
((member _X (_T . _Q)) (member _X _Q))
((conc () _X _X))
((conc (_T . _Q) _L (_T . _R)) (conc _Q _L _R))
((nrev () ()))
((nrev (_T . _Q) _R) (nrev _Q _Q1) (conc _Q1 (_T) _R))
((del _X (_X . _Q) _Q))
((del _X (_T . _Q) (_T . _R)) (del _X _Q _R))
((perm () ()))
((perm _L (_T . _Q)) (del _T _L _L1) (perm _L1 _Q))
```

The knowledge base is organized in sets of clauses distributed according to the head predicates (see section 2.1.2). The coding is done on the P-list of the predicate atoms under the `clauses` indicator. To add new clauses easily, let us define the macro-character `$`:

```
(set-macro-character
   #\$
   #'(lambda (stream char)
        (let ((cl (read stream)))
          (setf (get (caar cl) 'clauses)
                (append (get (caar cl) 'clauses) (list cl))))
      (values)))
```

Each time a new clause is read, it is added at the end of the set, itself coded as a list of clauses.

2.4.2. Renaming of Variables

Very often, the same clause is used in different ways in the proof of a goal (it is a clausal form of predicate calculus where the variables are universally quantified; the scope of a variable is therefore local to the clause where it appears).

To be able to describe without ambiguity the ties between variables and terms arising from the unification, it is necessary to duplicate, at each new use, the clause candidate for the unification [BCU82]. The duplication of a clause consists therefore in generating an image in which all the variables have been renamed.

To handle this problem, we choose to date the variables with the instant when the clause in which they appear is used, that is, following the chronology of the proof. The duplication is carried out by the function `rules`, which makes a copy of the set of clauses and renames the variables:

```
(defun rules (litt n)
   (rename (get (car litt) 'clauses) n))
(defun rename (x n)
   (cond
      ((variable? x) (cons x n))
      ((atom x) x)
      (t (cons (rename (car x) n) (rename (cdr x) n)))))
```

A variable thus becomes a dotted pair (*name.time_of_use*). Two forms appear for the same variable: the external one (its name) recognized by the primitive `variable?`, and the internal one (its coding) which will be recognized by the primitive `var?`:

```
(defconstant marque \#\_)
(defun variable? (x)
   (and (symbolp x) (char= (char (string x) 0) marque)))
(defun var? (x)
   (and (consp x) (variable? (car x)) (numberp (cdr x))))
```

Finally, let us note that our technique of duplication by systematically generating new copies is extremely expensive. We will return to this problem in the second part, where we will propose a much better solution (see section 4.2.1).

2.4.3. The Prolog Control

The Prolog control operates by means of two functions, solve and prove, which perform the forward and backward processes described in section 2.2.2.

First, solve analyzes a state of resolution described by a resolvent, a current environment, and the current time. The function solve distinguishes two cases:

- The resolvent is empty;
  we have therefore obtained a solution, and we must print the values of the variables appearing in the query.

- The resolvent is not empty;
  we start the proof of the first goal using the set of clauses that defines it. The state of the computation is then characterized by the tail of the current resolvent, the current environment, and the current time.

```
(defun solve (res env chr)
   (if (null res)
      (answer env)
      (prove (car res)
             (rules (car res) chr)
             (cdr res)
             env
             chr)))
```

Finally, let us note that systematically selecting the left-most goal conforms to the computation rule described in section 2.2.1

The function prove carries out the proof of a goal g using its set of clauses p, in a current state of resolution characterized by a resolvent r, an environment e, at time chr:

```
(defun prove (g p r e chr)
   (if (null p)
      'fail
      (let ((uni (unif g (caar p) e)))
         (if (eq uni 'fail)
            (prove g (cdr p) r e chr)
            (if (eq (solve (append (cdar p) r)
                           uni
                           (+ chr 1))
                    'fail)
               (prove g (cdr p) r e chr)))))))
```

If the set of clauses is not empty, prove tries to unify the goal with the head of the first clause. If unification fails, prove tries with the remaining clauses.

If it succeeds, then prove tries to solve the new resolvent consisting of the old one, at the head of which has been added the body of the clause satisfying the unification. The proof is then carried out in the environment uni at time 1+chr.

In case of failure, prove makes a new attempt for the current goal g with the remaining set of clauses. The proof begins with the initial state of the computation, that is the resolvent r, the environment e, and time chr.

Let us close with two comments:

- The clause and literal selection rules described in section 2.2.2 are satisfied; indeed, the clauses are considered in the order of the set and the left-most literal that is always selected first.

- The backtracking is done against time; indeed, when a failure occurs, the proof restarts at the corresponding level (last choice point) by trying the clauses that have not yet been considered.

2.4.4. Printing of Results

When a solution has been obtained, it is printed: if there are variables in the query, we provide their values; if not, then we merely answer in the affirmative. In the first case, the user is asked to look for another solution.

```
(defun inst (te s)
   (cond
      ((var? te)
         (if (bound? te s) (inst (ult te s) s) te))
      ((atom te) te)
      (t (cons (inst (car te) s) (inst (cdr te) s)))))
(defun answer (s)
   (if (null *lv)
      (format t "Yes ~%")
      (progn
         (mapc
            #'(lambda (x)
               (format t "~A = ~A~%" (car x) (inst x s)))
            *lv)
         (if (and (princ "More : ") (string= (read-line) ";"))
            'fail))))
```

*lv is a global variable containing the list of all variables appearing in the query. To print the value of a variable, we first construct the term to which it is bound. The search for a new solution requested by the user is provided by the function answer, which forces backtracking by simulating a failure.

2.4.5. The Interpreter

The interpreter loop reads the query, renames the variables to guarantee co-
herence with the duplication of the clauses, constructs the list of the variables
occurring in the query, then starts the proof (with the query as initial resolvent,
an empty environment, and the starting time 1).

```
(defun prolog () (myloop (myread)))
(defun myread ()
   (terpri)
   (format t "| ?- ")
   (read))
(defvar *lv)
(defun myloop (query)
   (if (eq query 'halt)
      'bye
      (let ((a (rename query 0)))
         (setq *lv (reverse (lvar a ())))
         (if (eq (solve a () 1) 'fail)
            (format t "No ~%"))
         (myloop (myread))))))
```

The exit from the interpretation loop is done by the atom halt. We still have
to construct the list of variables occurring in the query q; for this, we can use a
buffer variable buff:

```
(defun lvar (q buff)
   (cond
      ((var? q)
         (if (member q buff :test \#'equal) buff (cons q buff)))
      ((atom q) buff)
      (t (lvar (cdr q) (lvar (car q) buff))))))
```

2.4.6. The Unification

The implementation of unification described in section 1.5.2 must be slightly
modified. In fact, if the equality test done by eql is always correct for the
atomic constants, it is no longer valid for the variables represented by dotted
pairs. The test must therefore be done by equal.

```
(defun unif (x y s)
   (let ((x1 (val x s)) (y1 (val y s)))
      (cond
         ((equal x1 y1) s)
         ((var? x1) (add x1 y1 s))
         ((var? y1) (add y1 x1 s))
         ((or (atom x1) (atom y1)) 'fail)
         (t (let ((news (unif (car x1) (car y1) s)))
            (if (eq news 'fail)
               news
               (unif (cdr x1) (cdr y1) news))))))))
```

The primitives value and bound? described in section 1.5.2 must also be adjusted; add, val, and ult remain unchanged.

```
(defun value (x s)
   (cdr (assoc x s :test \#'equal)))
(defun bound? (x s)
   (assoc x s :test \#'equal))
(defun add (x val s)
   (acons x val s))
(defun val (te s)
   (if (var? te) (ult te s) te))
(defun ult (v s)
   (if (bound? v s) (val (value v s) s) v))
```

We have completely described the code for our pure Prolog interpreter. For other versions of Prolog microinterpreters in Lisp, we refer the reader to [Per82], [Nil84], [RK85], and [Fer86].

Finally, we close this chapter by noting that the Prolog control that we have defined is quite rudimentary [Kow80] because it makes a depth-first traversal from left to right of the search tree. Indeed, in the forward process, it systematically selects the first literal as well as the first clause. Then, in case of failure, the backtracking is always directed toward the last choice point. Such a strategy is subject to loops and may involve the needless construction of certain subtrees of the search tree.

CHAPTER 3

IMPROVING PROLOG CONTROL

Different approaches have been proposed to improve the classical strategy of Prolog control. H. Gallaire and C. Lasserre [GL82] suggest classifying control according to four principal levels.

1. Pragmatic control takes advantage of the depth-first left-to-right traversal strategy for writing Prolog programs.

2. Explicit control incorporated in the program consists of modifying the standard control strategy by means of built-in predicates or variables annotation. Several Prolog systems offer such possibilities: let us mention Prolog-II [Col82b], LM-Prolog [CK83], SP_Prolog [SPP87], SICStus-Prolog [Car87], IC-Prolog [CCG82], MU-Prolog [Nai85a], and NU-Prolog [Nai86], [Nai88].

3. Explicit control separated from the program differs from the above by the fact that the desired control is now expressed in the form of rules separate from the program. Here we see emerging two levels of knowledge: the program itself (facts and rules) and the way to execute it (metarules) [Din79], [Din80], [GL82], [BK82], and [Per84b].

4. Automatic control is also characterized by introducing strategies aimed at improving the efficiency of the control. But, in contrast to the preceding levels, these improvements are carried out systematically by the system itself. The best-known example is certainly that of the intelligent backtracking [LG80], [BP84], [Cox84], [CCF86], and [CCF88].

After first discussing these four levels of control, we will pay special attention to the explicit control incorporated in the program. We will restrict our study to three propositions: first, the variables annotation of IC-Prolog [CCG82], then the *wait declarations* of MU-Prolog [Nai85a], [Nai85b], and finally, the *freeze* predicate of Prolog-II [Col82b]. For each of these three propositions we will use the same example, *same_leaves*. We will then compare the use of the *wait declarations* with the predicate *freeze*. We will conclude with a brief survey of the *when declarations* of NU-Prolog.

3.1. Different Levels of Prolog Control

3.1.1. Pragmatic Control

The pragmatic control depends on the ability of the programmer to use the fixed strategy of control, for example, by ordering clauses and literals in order to obtain the desired result. This approach often has the unfortunate effect of destroying the initial declarative aspect of the solution. E. Elcock has very critical thoughts on this subject in [Elc83].

We will not discuss pragmatic control any further. We recommend the following introductory texts to Prolog programming: [CCP80], [CM84], [CC84], and [Diz89], as well as books also treating applications of Prolog, such as [KS85], [GKPC86], [Con86], [Bra86], [SS87], [Mal87], [WCS87], and [GLD89].

3.1.2. Explicit Control Incorporated in the Program

The explicit control incorporated in the program gives the programmer the ability to modify the standard control, either by means of built-in predicates or variables annotation.

A first form of explicit control included in the program is the use of Prolog built-in predicates for control. Unfortunately, these are in general far too few and rather rudimentary. The best known among them is the *cut*, which allows the pruning of the *and/or* search tree. We will discuss its role in section 14.3.1.

A more elaborate form of explicit control incorporated in the program is the use of mechanisms allowing synchronization of the execution of various goals, either by annotating variables or by using delaying mechanisms. This form of control will particularly interest us in this chapter.

To this end, we will study three proposals:

1. the variables annotation of IC-Prolog, which allows for a data-flow control (see section 3.2);

2. the *wait declarations* of MU-Prolog, where the execution of a goal is made subordinate to conditions on the unification of its arguments (see section 3.3); and

3. the *freeze* predicate of Prolog-II, which delays the proof of a goal as long as a variable remains free (see section 3.4).

To compare these three proposals, we will consider the problem of determining if two binary trees have the same leaves. For this, we let $a(g,d)$ denote the tree with left child g and right child d, whereas $f(b)$ will denote the tree reduced to a leaf b.

Trying to determine if two trees have the same leaves leads to the following program:

```
same_leaves(X,Y) :- leaves(X,F), leaves(Y,F).

leaves(f(X),[X]).
leaves(a(f(X),Y),[X|Fy]) :- leaves(Y,Fy).
leaves(a(a(X,Y),Z),F) :- leaves(a(X,a(Y,Z)),F).
```

In this form, to verify that two trees X and Y have the same leaves, we construct the list F of all the leaves of the first tree X, and then we compare F with the leaves of Y. If the two trees have the same leaves, this strategy is not too costly since the two trees must be traversed in their entirety anyway. On the other hand, if they differ by a leaf f(Z), it is not necessary to traverse either one beyond f(Z). From this, we get the idea of testing the equality of leaves one step at a time by using a coroutining traversal of X and Y.

3.1.3. Explicit Control Separated from the Program

In this approach, the control is expressed outside the program, which preserves its initial declarative aspect. A second level of knowledge above that of the program itself appears here: the metalevel. This level is specified in the form of rules (metarules) usually written in Prolog [Din79], [Din80], [GL82], [BK82]. The separation of the program and its control makes it easier to change the behavior of the program. We can thus define new rules for literal and clause selection in order to obtain the control strategy best suited to the problem at hand.

For further details, we refer to [GL82] and [BK82] as well as to the study of the MetaLog system introduced by M. Dincbas [Din79], [Din80].

3.1.4. An Example of Automatic Control: Intelligent Backtracking

Prolog backtracking systematically returns to the latest choice point. This way of proceeding is, in some cases, inefficient since it "blindly" reconsiders the choices made up to that point. Intelligent backtracking addresses this problem. It avoids the unnecessary traversal of subtrees, which can only produce failures, and thus improves the system performance. Intelligent backtracking analyzes the causes of failure in a unification in order to determine the binding or bindings that produced the failure.

The proposals of M. Bruynooghe and L. M. Pereira [Bru81], [BP81], [BP84], [PP82] are based on the search for minimal subtrees on which unification is impossible. P. Cox and T. Pietrzykowski [CP81], [Cox84], [PM83] use deduction tracings that are, in some respects, a generalization of the *and/or* proof tree to a graph.

According to M. Bruynooghe and L. M. Pereira [BP84], these two approaches are complementary. Indeed the solution proposed by P. Cox and T. Pietrzykowski [Cox84] amounts to searching the maximal subtrees on which unification is possible. We refer the interested reader to these two articles as well as to [CCF86], [CCF88].

Unfortunately these interesting ideas have not yet directly influenced the existing implementations. In fact, the various algorithms proposed until now are relatively complex and their applications too costly relative to the gains achieved.

For a complete picture of the different forms of Prolog control, we refer the reader to [GL82], as well as [LL86] where another classification can be found depending on whether the control is (mixed approach) or is not (separate approach) incorporated in the program.

3.2. Variables Annotation in IC-Prolog

IC-Prolog [CC79], [CC80], [CCG82] uses a mechanism of variables annotation that enables the data flow to direct the control. The order in which goals are proved is managed by the information flow within the variables shared by the goals.

In IC-Prolog one can specify if a goal B containing a variable X will behave as a producer or consumer of the variable X; that is, if it must give a value to X (producer), or if it must only use the value that others have given, or will give, to X (consumer).

All these conditions are specified in the definition of the predicate using a mechanism of variables annotation. There are two kinds of annotations associated with the notions of producer and consumer.

3.2.1. Lazy Producers

An annotation of lazy producer on a variable X (denoted by $X^\wedge$) in a goal B:

1. gives the goal the lowest priority,

2. prevents another goal from binding this variable;

that is, when another goal G tries to bind the annotated variable X to a term te (nonvariable), G is delayed, B is executed until it binds X, then G is executed.

The effect is inherited; that is, if B has bound the variable X to a term te, B will be activated as soon as another goal attempts to increase the degree of instantiation of te.

B is therefore a lazy producer with respect to the annotated variable X; that is, it will not instantiate it unless requested to do so by the others, particularly when the others are ready to instantiate it in its place.

3.2.2. Eager Consumers

An annotation of eager consumer on a variable X (denoted X?) in a goal B:

1. gives the goal the highest priority,

2. makes it the first one to consume the value of X;

that is, if during execution another goal G binds the annotated variable X to a term te (nonvariable), the current execution is suspended, B is proved as long as the degree of instantiation of X is unchanged, then the current execution continues.

The effect is inherited; that is, if X has been bound to a term te, B will be activated as soon as the degree of instantiation of te is changed. B is therefore an eager consumer with respect to the annotated variable X; that is, it uses (consumes) its value immediately with the evolution of its instantiation.

3.2.3. same_leaves in IC-Prolog

The desired solution to *same_leaves*/2 may be obtained by specifying a relationship of producer-consumer on the variable F shared by the two goals *leaves*(X, F) and *leaves*(Y, F) in the definition of the predicate *same_leaves*/2.

As quoted by K. Clark in [CCG82], two solutions are possible: to give F, in the second goal, the status of either lazy producer or eager consumer. Let us first force the second call to be a lazy producer of F, that is, it will not produce a new leaf for comparison except when asked by the first:

```
same_leaves(X,Y) :- leaves(X,F),
    leaves(Y,F^).

leaves(f(X),[X]).
leaves(a(f(X),Y),[X|Fy]) :- leaves(Y,Fy).
leaves(a(a(X,Y),Z),F) :- leaves(a(X,a(Y,Z)),F).
```

Otherwise, let us make the second goal an eager consumer of F so that it will attempt verification each time the first one generates a leaf:

```
same_leaves(X,Y) :- leaves(X,F),
    leaves(Y,F?).
```

We are thus led to two solutions of the given problem. In the first version, the second call produces and the first one consumes. In the second version the situation is reversed because each new leaf found by the first call is immediately consumed by the second. To learn more about the features of IC-Prolog, especially the parallel evaluation of goals, we recommend [CCG82].

3.3. The *Wait Declarations* of MU-Prolog

MU-Prolog [Nai85a], [Nai85b] uses the *wait declarations* in order to delay goals as long as they do not satisfy a certain number of conditions on the unification of their arguments. In this way the proof of a goal depends on the degree of instantiation of its arguments. We will first introduce the *wait declarations* as well as an example followed by the MU-Prolog's solution to *same_leaves*/2.

3.3.1. The *Wait Declarations*

The *wait declarations* define in a static manner the activation mode of each call G to a predicate p by specifying conditions on the unification of the arguments of G. When a goal G is called, the control examines these conditions. If they are satisfied, then G is immediately proved. Otherwise G is delayed until they are.

The activation conditions of a goal are defined relative to the construction of its arguments during the unification with a head clause having the same predicate symbol. An argument arg is said to be built by a unification if and only if there exists a variable X in arg that is bound by this unification. To make these declarations, MU-Prolog uses n-tuples consisting of zeros and ones, where the ones indicate the rank of the arguments that can be built.

Let us look at the example given by L. Naish in [Nai85a] concerning the concatenation of two lists:

```
| ?- listing(conc).
  conc([],X,X)
  conc([T|Q],L,[T|R]) :- conc(Q,L,R)
| ?- wait  conc(1,1,0).
| ?- wait  conc(0,1,1).
```

The two *wait declarations* stipulate that every call to conc/3 will be executed at once if it builds its first and/or second argument or its second and/or third argument. Here it is easier to characterize the delaying conditions: a call to conc/3 will be delayed if it tries to build its first and third arguments simultaneously (which corresponds to the well-known case of a loop).

So, if a goal B does not satisfy its *wait declarations* it is delayed (the degree of instantiation of its arguments is still too low). B will be awakened as soon as, in the rest of the proof, another goal G binds one or more of its variables, and in this way B satisfies the *wait declarations*.

3.3.2. Example

Let us consider L. Naish's example of the use of conc/3 with *wait declarations* for the concatenation of three lists [Nai85a]:

```
conc3(A,B,D,E) :- conc(A,B,C),
   conc(C,D,E).
```

First, without *wait declarations*, there is no problem concatenating three lists. On the other hand, the decomposition gives rise to a loop:

```
| ?- conc3([1], [2,3], [4], X).
   X=[1,2,3,4]
| ?- conc3(X, [3], [4], [1,2,3,4]).
   X=[1,2]
```

The subgoal conc(X, [3] ,C) will successively generate all the lists of the form concatenation of X (general model of list) and [3], i.e C=[3], C=[X1,3], C=[X2,X1,3], which leads to a success, and then continues to generate, entering into an infinite loop. Indeed, all the lists C produced after the first success will have length greater than or equal to four; so that conc(C, [4], [1,2,3,4]) will always lead to a failure for such values of C.

This use of conc3/4 behaves proportionally to the square of the length of the list we are seeking and enters an infinite loop. Note that if we reverse the order of the two calls to conc/3, a similar problem will occur during concatenation.

Let us look at the same example, but now including the *wait declarations* for the predicate conc/3 as defined in section 3.3.1. Indeed such a behavior of conc3/4 is related to the calls to conc/3 whose first and third arguments both are not instantiated.

The first call to conc/3 must obey the *wait declarations*; that is, it must wait for the second call to instantiate C before advancing another step and then wait again and so on until the solution is found, as shown in Figure 3.1.

Each recursive call to conc/3, generated by the proof of the first goal (on the left in the diagram) waits before proceeding until the corresponding call arising from the proof of the second goal (on the right in the diagram) instantiates its third argument. Control then alternates between the left branch and the right branch throughout the construction of C. With respect to this variable, the first goal behaves like a consumer and the second one like a producer.

The execution of conc3/4 becomes linear in the length of the list to be calculated and no longer enters a loop. The first goal turns into a consumer and no longer blindly produces an infinite number of solutions.

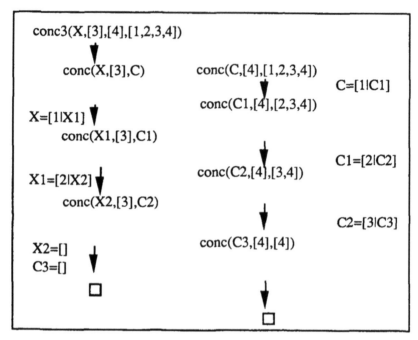

FIG. 3.1. *conc*

3.3.3. *Same leaves* in MU-Prolog

Since the two calls to `leaves/2` in the definition of `same_leaves/2` are symmetrical, it seems difficult to obtain the desired result except by using a trick to activate them.

Indeed, if a *wait declaration* is used to specify their behavior, it will be the same for both; either they will both be executed (and the first one entirely before the second) or else both will be delayed.

We therefore introduce a triggering mechanism (predicate `islist/1`), which will control the order of execution of the different goals by forcing the instantiation of the shared variable:

```
same_leaves(X,Y) :- leaves(X,F),
    leaves(Y,F),
    islist(F).
leaves(f(X),[X]).
leaves(a(f(X),Y),[X|Fy]) :- leaves(Y,Fy).
leaves(a(a(X,Y),Z),F) :- leaves(a(X,a(Y,Z)),F).

islist([]).
islist([_ | Q]) :- islist(Q).
```

All we have to do now is to make sure that every call to `leaves/2` waits until its second argument is instantiated, by declaring `|?-wait leaves(1,0)`.

The initial call to `same_leaves/2` leads to the delaying of the two goals `leaves(X,F)` and `leaves(Y,F)` until `islist/1` acts. Then, the first goal looks for its left-most leaf and submits it to the second for verification and delays. The second one verifies and also delays. Then `islist/1` is activated once again. This process continues as long as the leaves considered are identical and the two trees have not been completely traversed.

3.4. The Predicate *freeze*/2 of Prolog-II

Prolog-II [Col82b], [CKC83], [Can84], [Can86] has introduced the predicate *freeze*/2 in order to delay the proof of a goal for as long as a variable remains free. This delaying mechanism is also available in LM-Prolog [CK83], Prolog-Criss [Man85], SP-Prolog [SPP87], and SICStus-Prolog [Car87].

We will look at the predicate *freeze*/2 by means of examples. We will then study Prolog-II's solution to the problem of *same_leaves*/2. We will conclude with a comparison between the use of the predicate *freeze*/2 and the use of the *wait declarations*.

3.4.1. The Predicate *freeze*/2

The predicate *freeze*/2 is used in order to delay the proof of a goal B so long as a variable X remains free: when *freeze*(X, B) is called, if X is bound to a constant or a structured term, B is immediately proved. Otherwise B is delayed. B will be executed as soon as another goal G binds X to a nonvariable term during the proof.

Let us consider two examples of the use of the predicate *freeze*/2. The first one reflects a producer-consumer scheme, the second one expresses the use of constraints for the construction of lists.

3.4.1.1. A Producer-Consumer Scheme

The predicate *freeze*/2 allows us to introduce a producer-consumer scheme. Information is transferred by means of one or more shared variables. These variables are usually lists that are not completely instantiated.

The producer builds such a list recursively throughout the calculation. As soon as a new element is obtained, the consumer uses it immediately and then delays. At the beginning, we must freeze the consumer, and then activate the producer.

Let us look at an example of communication between two processes based on [TF83]:

```
test :- freeze(V,consom(V)),
   product(0,V).

product(X, [X|L]) :- product(X,X1),
   product(X1,L).
product(X,X1) :- plus(X,1,X1).

consom([X|L]) :- proceed(X),
   freeze(L,consom(L)).
proceed(X) :- write(X).
```

The first process puts out an integer at each new generation. The second one prints the integer it receives each time. The process starts with an initial value of zero. The shared variable V is used as a channel of communication between the producer and the consumer.

3.4.1.2. Constraints for List Construction

Another application of the predicate *freeze*/2 is the use of constraints for list construction. The constraint is initially frozen on the free variable representing the list to be built. The constraint will be verified each time a new element is added and reproduced recursively on the tail of the list still to be built.

We can also look at this problem as a producer-consumer relationship: the builder of the list is the producer and the constraint is a special kind of consumer whose unique aim is to validate or reject the proposed solution.

Let L be the language defined by the following grammar: $[V_t = \{a, b, c\}, V_n = \{S, S_1, S_2\}, S, R = \{S \rightarrow S_1 S_2; S_1 \rightarrow \emptyset | a S_1 b; S_2 \rightarrow \emptyset | c S_2\}]$ where $\emptyset$ represents the empty word.

Let us look for all the words of the language L of length less than or equal to four ($L = \{a^n b^n c^p; n, p \geq 0\}$). We will represent words by lists and write an acceptor of the language by associating a clause to each grammar rule [GKPC86], [Con86]:

```
word(X) :- s(X,[]).
s(X,Y) :- s1(X,Z), s2(Z,Y).
s1(X,X).
s1([a|X],Y) :- s1(X,[b|Y]).
s2(X,X).
s2([c|X],Y) :- s2(X,Y).
```

Next, let us define the predicate lg_inf/2, which checks if a list L is of length less than or equal to a natural number N (here we are using the representation of integers described in section 2.3.3).

```
lg_inf([], _).
lg_inf([_ | Q], s(X)) :- lg_inf(Q,X).
```

Since the acceptor can also be used as a generator, let us look for the words in the language of length less than or equal to four. The interpreter will then produce a first set of answers and will enter a loop.

```
| ?- word(X), lg_inf(X,s(s(s(s(zero))))).

   X=[]
   X=[c]
   X=[c,c]
   X=[c,c,c]
   X=[c,c,c,c]
```

Each time a call to `lg_inf/2` fails, backtracking comes back to the latest choice point in s2/2, in order to select the second clause. But, as s2/2 is recursively defined, control enters a loop. All the words derivable by using the second clause defining s1/2 are ignored.

This undesirable phenomenon occurs because the words of the language are generated before being tested (the constraint acts as a filter after the fact but does not influence the construction of the word).

Let us require the verification of the constraint during the construction of the list, by delaying, when necessary, the calls to `lg_inf/2`:

```
lg_inf([], _).
lg_inf([_ | Q], s(X))  :- freeze(Q, lg_inf(Q,X)).
```

Let us rephrase the query by freezing the constraint before proceeding with the generation. This time we do get the expected result:

```
| ?- freeze(X, lg_inf(X,s(s(s(s(zero)))))), word(X).
   X=[]
   X=[c]
   X=[c,c]
   X=[c,c,c]
   X=[c,c,c,c]
   X=[a,b]
   X=[a,b,c]
   X=[a,b,c,c]
   X=[a,a,b,b]
| ?-
```

Henceforth the interpreter produces all the solutions. Indeed, the second clause defining s2/2 will lead to a failure when it tries to generate a word of length five because of the constraint activation. By backtracking, control returns to the choice point in s1/2 in order to activate the second clause. The words derivable by s1/2 (other than the empty word) will this time be generated.

This example, taken from the definition of the predicate *liste − de − un/1* [GKPC86], is typical of the use of *freeze/2* in the construction of lists with constraints. For other examples, we refer the reader to [Kan82] and [GKPC86] as well as to [CH87].

3.4.2. *Same_leaves* in **Prolog-II**

As in MU-Prolog, we use the predicate `islist/1` to synchronize the traversals of the two trees. We must therefore delay the two initial calls to `leaves/2` then force the instantiation of the shared variable F in order to start them up again:

```
same_leaves(X,Y) :- freeze(F,leaves(X,F)),
    freeze(F,leaves(Y,F)),
    islist(F).
leaves(f(X),[X]).
leaves(a(f(X),Y),[X|Fy]) :- freeze(Fy,leaves(Y,Fy)).
leaves(a(a(X,Y),Z),F) :- leaves(a(X,a(Y,Z)),F).

islist([]).
islist([_ | Q]) :- islist(Q).
```

A call to `leaves/2` must run down the tree recursively until the first leaf is found and then must wait. To do this, we freeze the recursive call to `leaves/2` in the second clause (Fy is a free variable at this time and the potential list of the leaves still to be traversed). On the other hand, the recursive call in the third clause is not delayed since it is merely a step in the search for the next leaf.

3.5. Comparison of *Wait* and *Freeze*

Three main characteristics differentiate the use of the *wait declarations* and that of the predicate *freeze*/2. First of all, the *wait declarations* pertain to the definition of a predicate and will therefore be taken into consideration at each call, whereas *freeze*/2 applies only to a specific goal. Next, a *wait declaration* involves several arguments, whereas *freeze*/2 can only involve a single variable. Finally, *wait declarations* are based on the notion of argument building whereas *freeze*/2 is based only on whether a variable is free or bound.

3.5.1. Each Call versus Specific Call

The *wait declarations* are related to the definition of a predicate. They will therefore be automatically taken into account each time there is a call to a goal with this predicate name. All the calls will thus have an identical behavior.

On the other hand, *freeze*/2 applies only to a specific goal. If several calls associated with the same predicate must be delayed, they must all be frozen separately as is shown by the preceding definition of `same_leaves/2` (see section 3.4.2). Generally, we define two levels of predicates, the first one being a "frozen version" of the second.

Let us redefine same_leaves/2 on this model:

```
same_leaves(X,Y) :- leaves(X,F),
    leaves(Y,F),
    islist(F).
leaves(X,F) :- freeze(F,leaves1(X,F)).

leaves1(f(X),[X]).
leaves1(a(f(X),Y),[X|Fy]) :- leaves(Y,Fy).
leaves1(a(a(X,Y),Z),F) :- leaves1(a(X,a(Y,Z)),F).

islist([]).
islist([_ | Q]) :- islist(Q).
```

Despite the disadvantage of the duplicate definition, this technique differentiates explicitly between the calls that must be delayed (leaves/2) and those that must execute normally (leaves1/2), unlike the case of *wait declarations*.

3.5.2. Arguments versus a Single Variable

The *wait declarations* specify the conditions for activating (and also for delaying) a goal containing one or more variables. *freeze/2*, on the other hand, applies only to one variable.

Nevertheless, using *freeze/2*, we can express multiple conditions on the activation of a goal B, whether it consists of disjunctions (to prove B, wait until at least one of the variables is bound) or conjunctions (wait until all variables are bound before activating B).

The implementation of a disjunction of conditions on a goal B is carried out by successively freezing B on each of the variables. This is sufficient to guarantee the delay. But, upon activation, a same goal may be proved as many times as there are frozen variables.

The solution [GKPC86], [Nai85b] consists in forcing all the instances of B to share a same variable. The first occurrence of B to be activated will bind this variable, in a sense preventing all attempts to start the proof of the others.

Let us define the predicate freeze_or2/3, which restrains the execution of a goal B to the binding of at least one variable between the two.

```
freeze_or2(X,Y,B) :- freeze(X, once(B,V)),
    freeze(Y, once(B,V)).
once(B,V) :- var(V),
    call(B),
    eq(V,done).
once(_,V) :- nonvar(V).
```

The predicate once/2 tests the instantiation of the shared variable: if it is the first activation (V is a free variable) then B is proved and the variable is bound; otherwise there is nothing to be done.

We can now suggest a solution to the behavior of conc3/4 as described in section 3.3.2.

```
conc3(A,B,D,E) :- conc(A,B,C), conc(C,D,E).
conc(X,Y,Z) :- freeze_or2(X,Z,conc1(X,Y,Z)).

conc1([],X,X).
conc1([T|Q],L,[T|R]) :- conc(Q,L,R).
```

Finally, we can generalize this process to an arbitrary number n of variables given in the form of a list. To do this, it is enough to reproduce the freeze recursively, and to make sure that, on awakening, all the instances of the goal share the same variable:

```
freeze_orn(L,B) :-
   freeze_n(L,B,V).
freeze_n([],B,V).
freeze_n ([T|Q], B, V) :- freeze(T,once(B,V)),
   freeze_n (Q,B,V).
```

A conjunction of conditions on a goal B can be implemented by successively interleaving *freeze* on the variables appearing in the conditions.

We define the predicate freeze_and2/3 that executes the goal B when the variables X and Y are both bound. We deduce the definition of the predicate sum/3, which executes only when its first two arguments are instantiated:

```
freeze_and2(X,Y,B) :-
   freeze(X, freeze(Y, B)).
sum(X,Y,Z) :-
   freeze_and2(X,Y,plus(X,Y,Z)).
```

More complex conditions can now be expressed either directly [GKPC86], [Gre85] or by a combination of freeze_and2/3 and freeze_orn/2.

3.5.3. Building an Argument versus Binding a Variable

The *wait declarations* are related to the notion of building an argument during unification, whereas the predicate *freeze*/2 applies to a free variable. Recall that an argument is said to be built by a unification if and only if there exists a variable of this argument that is bound by the unification.

Consequently, a goal with a functional argument that has not been completely instantiated may be delayed using a *wait declaration* but not by the predicate *freeze*/2. Let us look at the following example taken from [Nai85b]:

```
even(0).
even(s(s(X)))  :-  even(X).
```

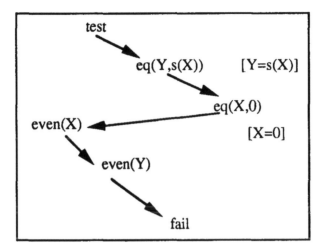

FIG. 3.2. Test

```
test  :-  even(X),
   even(Y),
   eq(Y,s(X)),
   eq(X,0).
```

Even integers are defined using the representation as successors of zero. We wish to answer the query test/0. In this form the program will enter a loop since it is necessary to delay the two calls to even(X) and even(Y).

Let us first examine (Figure 3.2) MU-Prolog's solution by declaring | ?- wait even(0). Every goal even(arg) that is building its argument will be delayed. The execution of test/0 leads, of course, to a failure.

Indeed, the two goals even(X) and even(Y) are initially delayed. Y is then bound to s(X) but the goal even(Y) is not yet activated because this would lead to building its argument (X would be bound to a term s(X1)). X is bound to zero and even(X) is activated and succeeds. As X is bound, we try to prove even(Y) (its unification does not build anything more), which leads to a failure.

Let us look now at the solution given by Prolog-II by means of the predicate *freeze*/2. We redefine two levels in order to have a systematic freeze on every call:

```
even(X)  :- freeze(X,even1(X)).
even1(0).
even1(s(s(X)))  :-  even(X)

test  :-  even(X), even(Y), eq(Y,s(X)), eq(X,0).
```

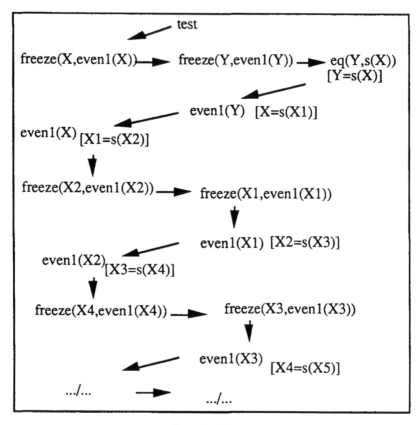

FIG. 3.3. Freeze

The proof of test/0 enters a loop. Indeed, the two initial calls to even/1 are, as before, delayed; but the goal even(Y) is reactivated right after the unification between Y and s(X).

This premature activation will trigger the proof of even(X), and these two goals will call each other indefinitely as shown in Figure 3.3 where all calls to the predicate even/1 have been directly replaced by their equivalent expression in terms of freeze/2.

First, eq(Y,s(X)) binds Y to s(X). even1(Y) is activated and binds X to the term s(X1), and the recursive call even1(X1) is delayed. Since X is bound, even1(X) is activated, and X1 is bound to s(X2) and even1(X2) is delayed. Control returns to even1(X1), which binds the variable X2 to the term s(X3), and even1(X3) is delayed. But, since X2 is bound, even1(X2) is activated and binds X3 to the term s(X4), and even1(X4) is delayed. Control returns to even1(X3), which . . .

Control thus switches alternately from one side to the other and unification adds, at each step, a new functional level to the variable Y (Y=s(X), X=s(X1), X1=s(X2), X2=s(X3) ...).

The loop is thus caused by the unexpected activation of even(Y) right after the binding Y=s(X). This phenomenon is due to the fact that it is impossible to impose a freeze on a structured term.

3.5.4. Implementing *freeze*/2 Using *Wait Declarations*

Finally, we consider the implementation of the predicated freeze/2 in MU-Prolog. We prevent the building of its first argument by using a *wait declaration*:

```
freeze(zzzzzz,_)  :-  fail.
freeze(_,B)  :-  call(B).

| ?- wait freeze(0,1).
```

The first clause defining freeze/2 is just a trick to guarantee holding. Indeed, without this clause, a goal freeze(V,B) cannot be delayed even if V is a free variable. The first argument V will never be built since it is always unified with a free variable. Therefore the first clause is added in order to force the building.

- If V is a free variable, the head of the first clause satisfies the unification (V=zzzzzz). Since its first argument is built, the goal freeze(V,B) is delayed. As soon as V is bound, freeze(V,B) is activated and the first clause leads to a failure. The second clause is then selected and B is then proved.

- If V is bound, the unification with the head of the first clause fails. B is called by selecting the second clause.

We thus find in both cases the usual behavior of the predicate freeze/2.

3.6. The *When Declarations* of NU-Prolog

The delaying mechanism of the *wait declarations* is used and extended by NU-Prolog [Nai86], [Nai88]. The activation conditions (and, by the same token, the delaying ones) of a goal are still defined statically but this time with the help of the *when declarations*.

The *when declarations* associated with the definition of a predicate determine the behavior of all its calls. A *when declaration* specifies the degree of instantiation that the arguments of a goal must have in order for the goal to be proven as soon as it is called; we can, for example, force a variable to be instantiated or to be bound to a closed term.

Let us look again at the examples of section 3.3: every call to `conc/3` must
be delayed if its first and third arguments are not instantiated; similarly for
`leaves/2` with its second argument.

```
| ?- conc(A,B,C) when A or C.
| ?- leaves(X,F) when F.
```

The *when declarations* allow specification of finer conditions than the ones
permitted by the *wait declarations*. Consider the partition of a list of integers
with respect to a given value N. Classically, one wishes to wait when the list, or
N itself, is unknown; but we could also force the head of the list as well as N to
be closed terms so that the tests done by the two predicates `le/2` and `lt/2` do
not produce errors.

```
partition( ([],_, [], []).
partition( [T | L],N, [T | L1],L2) :- le(T,N),
    partition(L,N,L1,L2).
partition( [T | L],N, L1, [T | L2] ) :- lt(N,T),
    partition(L,N,L1,L2).

| ?- partition([],_,_,_) when ever.
| ?- partition([T|_],N,_,_) when ground(T) and ground(N).
```

Thus, any call to `partition/4` having the empty list as its first argument
will normally be proved. The others must be delayed until the head of the first
argument as well as the second argument are known. To learn more about the
when declarations and the other characteristics of NU-Prolog, we recommend
[TZ86] and [Nai86].

Part II

Principles and Techniques of Implementation

The purpose of this second part is, first of all, to introduce the different problems encountered in implementing a Prolog system, then to describe and compare the various solutions that have been proposed. We will also give a brief historical overview and make an attempt at classifying the existing Prolog systems based on their main implementation characteristics, such as architecture of the working area, representation of terms, memory management, and extra control facilities.

Our presentation will follow the two levels of progression given in the introduction:

1. statement of the problems and the solutions in an interpreted approach leading to the compilation stage, and

2. transition from the implementation of a *classical Prolog* to that of a *Prolog+* with delay mechanism.

To begin, we will be interested in the first aspect, presenting, in succession, the critical points of the implementation of *classical Prolog*: backtracking, representation of terms, and memory management. Then, we will discuss *Prolog+*, concentrating on one of the three propositions introduced in chapter 3: the delay mechanism associated with the predicate *freeze* of Prolog-II.

From Interpretation to Compilation of Classical Prolog

The transition will be made in two stages: first, we will discuss, from an interpretative point of view, the specific problems relating to the implementation of Prolog, describe the different solutions, and compare them. We will then have all the elements necessary to tackle the question of compilation.

Step 1: Statement of the Problems, Solutions, and Comparison

The general idea here is to take advantage of the techniques of stack management already developed in other languages and use them for implementing Prolog. Of course, will have to adjust the usual techniques in order to solve the specific problems arising from unification and backtracking.

Furthermore, as the Prolog language requires a large memory space, we will pay particular attention to optimizing the memory management.

We will distinguish two levels of implementation, depending on the power of the memory recovery mechanisms used:

1. update upon return from deterministic calls, and

2. last-call optimization.

The first stage is achieved by considering the implementation of the control by stack management, the different modes of representation of terms, and the specific problems encountered by updating upon return from deterministic calls. The second stage will be obtained from the first by taking into consideration the requirements inherent to the last-call optimization.

- First stage: update upon return from deterministic calls

    - We will make the starting point explicit by comparing Prolog with a classical language (see chapter 4). We will then be interested in the implementation of the control by means of a single stack by describing the form of the activation blocks and the implementation of backtracking. We will be faced with the problem of restoring environments. This will be resolved by introducing a second stack (the *trail*) into the working area.

    - Chapter 5 will be devoted to the representation of structured terms in Prolog. We will first study the binding mechanism arising from the unification. We will then describe the two modes of representation of terms used in Prolog: stucture sharing and structure copying.

    - The update upon return from deterministic calls will be the subject of chapter 6. We will look at the problems and their solutions in the context of structure sharing and structure copying. This will lead us to a common architecture with three stacks: *local stack*, *global stack*, and *trail*. We will close with a comparison of the two approaches.

- Second stage: last-call optimization

    - In chapter 7 we will be interested in implementing last-call optimization. We will first introduce the principle behind it. Then, starting with the previous proposition (chapter 6), we will adjust the form of the activation blocks to the new requirements arising from the implementation of such a mechanism.

Finally, to conclude the first step, we will note that the memory management previously considered is very closely tied with determinism. Therefore, we will study (chapter 8) an implementation technique used to reinforce the determinism naturally—namely, clause indexing.

Step 2: Compilation of *Classical Prolog*

The knowledge gained in the first step will allow us to tackle the compilation stage (chapter 9). As a basis for our study, we will choose the Abstract Machine of D. H. Warren [War83]. After first describing its choices and major directions, we will be interested in the specific questions of compilation.

From *Classical Prolog* to *Prolog+*

We will consider (chapter 10) only one of the three propositions introduced in the first part: the predicate *freeze* of Prolog-II. We will study the implementation of such a delay mechanism based on the solution offered by Prolog-II [Can84], [Can86]. This implementation will be based on the second stage (chapter 7) adjusted to the requirements of the predicates *dif* and *freeze*. This will lead us to a new architecture with four stacks, supporting at the same time the last-call optimization and the delay mechanism. This proposition will constitute the third and last stage.

Finally, let us note that each of the three stages defined in this way will be the subject of a particular implementation in Part 3, namely: Mini_Cprolog (see chapter 11) for the first stage, Mini_WAM (see chapter 12) for the second, and Mini_Prolog-II (see chapter 13) for the third and last stage.

CHAPTER 4

CONTROL AND STACK(S) MANAGEMENT

Prolog is above all a programming language and its control can therefore be considered, from a classical point of view, in terms of calls to and returns from procedures with parameter passing. But the usual model must be extended [Byr79], [Byr80] in order to take into account the particularities of unification and backtracking.

To implement Prolog, we can take advantage of classical techniques of stack management already developed for other languages. We must, of course, make adjustments to the usual processes in order to meet the specific requirements of Prolog, especially those of unification and backtracking.

Our approach will be first to try to implement the Prolog control with the help of one stack. We will then come up against a problem inherent to backtracking: the necessity of restoring certain environments in order to recover the correct state of the computation. This problem is usually solved by introducing an additional stack: the *trail*. This will lead us to a first architecture of the working area divided into two stacks, updated only upon backtracking.

To do this, we will first recall the notion of procedure in Prolog. Then we will describe the stack management and the activation blocks. Finally, we will study the problem of restoring the environments.

In order to focus our attention on the implementation of the control and the problems inherent to backtracking, the examples used in this chapter will involve only simple terms. The representation of structured terms as well as of the associated binding mechanism will be presented in the next chapter.

4.1. Prolog as a Programming Language

We will first briefly recall the notion of procedure in Prolog. We will then extend the call-return model used in classical languages to include backtracking. Finally, we will study the peculiarities of parameter passing via unification.

4.1.1. Procedures

A set of clauses associated with a predicate p constitutes many possible definitions of a procedure with the same name p. The body of each clause is the body (eventually empty) of the procedure. Each body of a clause consists in a sequence of calls to the different literals of which it is composed. Thus, in contrast to the situation in a classical programming language, the definition of

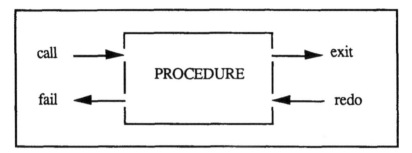

FIG. 4.1. Byrd model

a procedure will not be unique. Because of backtracking, the execution of a procedure can generate different results.

4.1.2. Control

The Prolog control is completely determined by the choice of a computation rule and a search rule (see section 2.2.2). The Prolog control can thus be interpreted in terms of the classical call-return model, with a new dimension added by the backtracking.

In fact, the computation rule entails the left-to-right traversal of the different *and* branches (see section 2.3.2), so we can associate it with the classical model in terms of procedure calls and returns: to execute a call, transmit the parameters, and execute the body of the selected clause sequentially—which for Prolog means to unify with the head of the clause and prove in succession all the goals constituting the body of the clause.

But backtracking introduces the notion of multiple successes and recalls. This new situation can be formalized by the model of the boxes of L. Byrd [Byr79], [Byr80], [Con86].

The execution of a goal is represented in the form of a box that can be entered or exited in four different ways (Figure 4.1). Let us describe each of the gates:

- *call* denotes the initial call,
- *exit* describes each successful exit,
- *redo* represents a new attempt at a proof, resulting from a failure, and
- *fail* denotes the final return once all possible proofs have been attempted.

But unlike the Byrd model, we will consider that the return from a call can result in either a success or a failure:

- If the call has produced a success without generating any intermediate choice points, we will say that the call is *deterministic* and *returns with success*.

- On the other hand, if all attempts have failed, we will say that the call *returns with failure*.

The return with success corresponds to the last success without intermediate choice points and the return with failure corresponds to the case of definitive failure.

This distinction aims at pursuing the analogy between the classical model of call-return (for which there would only be two gates) and the Prolog control. Indeed, when a call has produced a success without generating intermediate choice points, its execution is performed completely. One can then think about recovering the memory space needed for its execution as soon as it returns or as soon as the last goal of the selected clause is called.

We will present these two mechanisms for memory management respectively in chapter 6 and chapter 7.

4.1.3. Unification

We can compare the unification between a goal and the head of a clause to the passing of parameters between a calling procedure (goal) and a called procedure (head of clause).

But, unlike the usual parameter passing, the unification:

- establishes a communication in both directions and not uniquely from the caller to the callee;

- is not guaranteed to succeed and, in case of failure, does not produce errors in the usual sense but forces a backtracking;

- transmits terms that can be more or less instantiated;

- contains arguments that are not typed in general; and

- is not subject to distinction between input and output arguments; we can therefore in some cases reverse the roles and take advantage of the properties of reversibility.

In a compilation environment, parameter typing and specification of input and output arguments are used to optimize the unification and memory management [War77], [War83]. We will return to this first in section 6.4.3, where we will present the declaration mode of Prolog-Dec10 [War77], and in more detail in chapter 9, where we will specifically treat the various aspects of compilation.

4.2. Stack Management

We wish to implement the Prolog control by means of a single stack. At each call of a goal an activation block will be allocated on this control stack. But taking into account the backtracking, two kinds of activation blocks must be considered depending on whether they are associated with choice points. This stack will be updated only upon backtracking by popping all the blocks subsequent to the last choice point.

Each block will have a control part aimed at managing the forward and backtracking processes and an environment part arising from the duplication of the clause whose head has satisfied the unification.

After discussing the notion of environment in Prolog, we will describe the two kinds of activation blocks: deterministic blocks and choice blocks. We will conclude with an example.

4.2.1. Environments

The problem of multiple usage of the same clause (see section 2.4.2) is solved by generating a new exemplar of the clause at each use. This duplication [BCU82] is carried out by creating an environment that will receive the values of the variables appearing in the clause. The pair (clause model, environment) becomes the new exemplar of the clause. This process has the advantage that by automatically renaming the variables, it avoids copying the entire clause (see section 2.4.2).

Let cl be a clause involving n distinct variables. The variables are generally numbered 0 to $n-1$. The duplication of the clause cl produces the pair (cl, env), where env is an environment of size n that stores the values of the variables appearing in cl. Initially, these n variables are, of course, in a free state.

Example:
> The duplication of the rule $cl = conc([T|Q], L, [T|R]) : -conc(Q, L, R)$
> produces the pair (cl, Env) where cl is the model of the rule and
> Env is the new environment $[free, free, free, free]$. The variables
> T,Q,L,R are respectively numbered 0,1,2,3.

An environment is represented by means of a vector containing n memory locations. The value of a variable is obtained by using its range as an index into this vector.

The binding of a variable i to a term te is achieved by assigning to $Env[i]$ a reference to te. We will describe the binding mechanism as well as the different modes of representation of structured terms in chapter 5.

4.2.2. Activation Blocks

We have two kinds of activation blocks depending on whether they are associated with choice points: the deterministic blocks and the choice blocks. Each block contains a control part aimed at managing the forward and backtracking processes (see section 2.2.2) and an environment part arising from the duplication of the clause whose head has satisfied the unification.

4.2.3. Deterministic Blocks

When a goal b is called, a deterministic block is pushed on the control stack if the clause cl, whose head satisfies the unification, is the last clause in the set to be considered.

4.2.3.1. Composition of a Deterministic Block

Each deterministic block has two parts:

- the continuation part, which manages the forward process, is made up of two fields:
    - the field CP, which allows access to lb, the next sequence of goals to prove once the body of the clause cl has been proven,
    - the field CL, which indicates the activation block containing the information needed for the execution of lb;
- the environment part (vector of length n, number of variables occurring in cl), resulting from the duplication of cl.

The continuation part allows the location, during the forward process, of the next sequence of goals to prove (field CP) and the activation block (field CL) containing the environment in which the proof must be carried out.

4.2.3.2. The Continuation Part of a Deterministic Block

The continuation part of a deterministic block may indicate

- either the sequence (eventually empty) of sibling literals in the sense of the *and/or* proof tree (see section 2.3.2) and the associated block
- or directly the next sequence of goals to prove and the corresponding block.

The first choice completely respects the structure of the *and/or* proof tree (see section 2.3.2). In fact, for each block associated with a goal b, the field CP indicates the sequence of sibling literals of b, and the field CL indicates the block of the father goal of b (which contains the environment in which b and its siblings are to be considered).

But this choice forces, after the proof of the last literal of each clause, the traversal of the different Continuation fields in order to search for the next goal to prove. In the second choice, such access is privileged.

In this chapter and the two following it, we will consider the first approach. The second approach will be presented in chapter 7, which is devoted to the last-call optimization.

4.2.4. Choice Blocks

When a goal b is called, a choice block is allocated on the control stack if the clause cl, whose head satisfies the unification, is not the last clause in the set to be considered. Such a block is therefore a choice point.

4.2.4.1. Composition of a Choice Block

Each choice block has three parts:

- Rerun managing the backtracking, made up of three fields:
    - the field BL, referencing the previous choice block on the control stack,
    - the field BP indicating the remaining choices in the form of a set of clauses still to be considered, and
    - the field TR (we will describe its role in section 4.3);
- Continuation (as in the case of a deterministic block); and
- Environment (as in the case of a deterministic block).

A choice block can be distinguished from a deterministic block in that it contains, in addition, a part rerun aimed at implementing the backtracking.

The various choice blocks on the control stack are linked together via the field BL. Upon backtracking, it suffices to look at the register BL pointing to the last choice block in order to locate the restart point and to update the control stack. In case of repeated failures, this linking enables the successive execution of the different levels of backtracking.

The field BP is used during backtracking to restart the forward process by providing the candidate clauses for the unification of the goal under consideration. But nowhere have we saved the specific information needed to recover this goal.

4.2.4.2. Recovering the Goal to Restart the Forward Process

A simple solution to this problem is to use the continuation part of the block by making CP point to the goal itself followed by its sibling literals, rather than to the sequence of its sibling literals. Upon backtracking, a simple check with CP will locate the goal and its environment held by the father block pointed to by

CL. In this case, the backtracking will need information from the continuation part in addition to information from the rerun part.

Another solution would be to save the arguments of the goal in each choice block. We will consider, for the moment, the first approach. We will take up the second approach in chapter 7, which is devoted to the last-call optimization.

4.2.5. Meaning of the Various Fields

In summary, and according to the choices made previously, during the proof of the goal bi with siblings $bi + 1, \ldots, bn$, the meaning of the various fields of the activation block associated with bi are as follows:

- CP, which designates the sequence $bi, bi + 1, \ldots, bn,$
- CL, which indicates the block of the father goal on the control stack,
- the Environment part (vector of length n number of variables occurring in the clause satisfying the unification),
- the field BL, which points to the previous choice block on the control stack,
- the field BP, which indicates the remaining choices in the form of a set of clauses still to be considered,
- the field TR, whose role will be described in section 3.3.

The Rerun part consisting of the last three fields exists only in the case of a choice block.

4.2.6. Example

Let us look at an example restricted to propositional calculus (see Figure 4.2) in order to explore further the implementation of the control.

Let us analyze the state of the control stack associated with the proof of the goal p (see Figure 4.3). Since all environments have an empty size, they have been omitted.

We create an initial block associated with the query. Then for each call a block is allocated. Each activation block reduces to the two fields CL and CP except for the goal q, where a choice point has been allocated. The BL register designates the last choice point. Upon backtracking, it will allow the forward process to be restarted and the control stack to be updated.

The working area consists of, for the time being, a single stack updated only upon backtracking with the help of register BL that points to the last choice point. Access to the various fields of this block allows the forward process to be restarted.

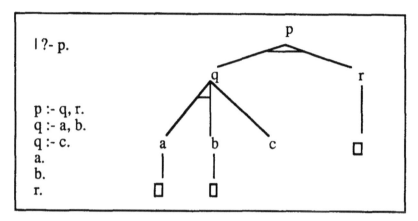

FIG. 4.2. The proof tree

This approach is correct if we can guarantee that the state of the stack underneath this block has not been altered since the creation of the choice point. Since the continuation and rerun parts are unchanged, there remains only the question of the environments.

4.3. Restoring the Environments

Taking into account the joint action of duplication and unification, we see that certain environments prior to the last choice point may have been irreparably modified since the creation of this choice point. Upon backtracking, simply updating the top of the stack does not guarantee in any way that we will recover the exact state of the bindings at the time the choice point was created. We will illustrate the problem with an example and then describe the solution proposed by the introduction of a second stack (the *trail*) into the working area of the Prolog system. This stack saves, during the proof, all the modifications brought about in the environments prior to the last choice point. Upon backtracking, it will then be possible to restore the modified environments to their initial states.

4.3.1. Postbindings

Since duplication results in representing the environments as globally accessible vectors, certain environments prior to the last choice point could have been modified by postbindings.

Let c be the instant when the last choice point has been allocated. Let xi be a free variable of an environment E allocated at time j such that $j < c$. xi is said to be postbound if and only if xi is bound at time $t \geq c$.

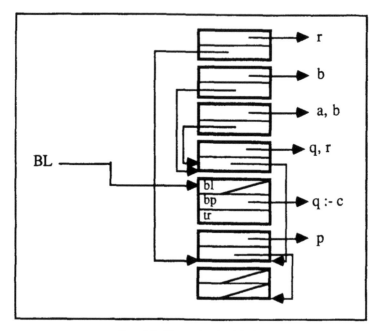

FIG. 4.3. The activation blocks

Let us look at an example showing the creation of a postbinding (see Figure 4.4)

The state of the control stack is that preceding the call to $r(Y)$. Three blocks have been allocated. The first one, associated with the query, has an empty environment. The second, associated with the call to $p(a)$, creates an environment of size 2 where X is bound to the constant a. The third, arising from the call $q(X, Y)$, generates a choice point with an empty environment. During the unification between $q(X, Y)$ and $q(a, b)$ the variable Y, whose creating time is prior to BL, is postbound to the constant b.

Since the goal $r(Y)$ fails, the control backtracks. The forward process is restarted via the register BL. But simply updating the top of the control stack does not permit the recovery of the previous state of the computation. In fact, Y remains bound to b while it was previously free.

4.3.2. The *Trail*

This problem is handled in all Prolog systems by introducing a new stack in the working area. This memory area is called *trail* [War77], [War80], [War83], [Bru76], [Bru80], [Mel80], or *restoring stack* [Can84], [Can86], or also *stack of assigned variables* [CKC79]. The *trail* is used to locate, during the forward

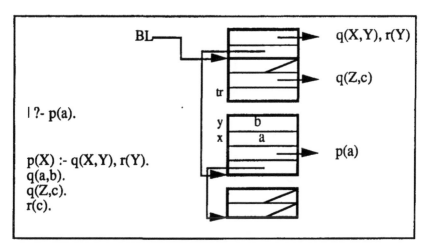

FIG. **4.4.** An example of postbinding

process, the variables that are subject to postbindings and to restore these variables to their initial value upon backtracking. After discussing the management of this stack we will study its use on the example of Figure 4.4.

4.3.2.1. Management of the *Trail*

During the forward process, any variable that is postbound is pushed on the *trail*. To do this, for each binding of a variable we compare its creating time with that of the last choice point. If it is earlier, the variable is saved on the *trail*.

Upon backtracking this stack is updated and, at the same time, the variables pushed since the creation of the last choice point are freed. For this purpose, each rerun part of a choice block has a field TR (see section 4.2.4) that saves the current top of the *trail*. Upon backtracking, this field sets the limit to which the updating must be carried out.

Let us specify the algorithm for restoring the environments. The top of the *trail* points to the first free space available on the stack.

begin
    let *top* be the current top of the trail;
    let *oldtop* be the value of the field TR of the last choice point;
    while (*top≠oldtop*) do
        top := *top* − 1;
        let *v* be the variable located at the *top* of the trail;
        reset *v* in a free state;
    endwhile;
end;

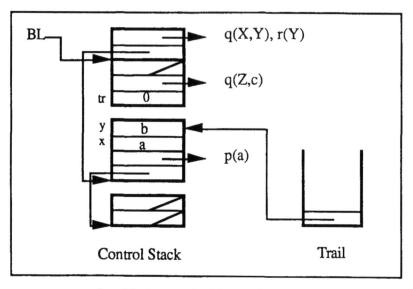

FIG. **4.5.** An example of the use of the *trail* (a)

Restoring the environments upon each backtracking guarantees that the exact state of the bindings is recovered and therefore that the proof can be restarted with the initial conditions.

4.3.2.2. Example

We consider again the example described in Figure 4.4 in order to illustrate the use of the *trail* (see Figure 4.5).

The field TR of the rerun part of the choice point indicates that at the time of its creation the *trail* was empty. Then, during the unification between $q(X, Y)$ and $q(a, b)$, Y is pushed on the *trail* because it is postbound to b.

Upon backtracking due to the failure of the goal $r(Y)$, environments are restored. The state of the two stacks is then as shown in Figure 4.6.

The choice point has disappeared since there is now only one clause to consider. Because of the restoration of the environments the variable Y recovers its free state.

Let us now complete the proof (see Figure 4.7). This time, the call to $q(X, Y)$ creates a deterministic block, and Z is bound to the value of X, and Y is bound to the constant c. But, in the absence of choice points, Y is not saved on the *trail* that remains empty. Then $r(Y)$ is executed, securing the final success of $p(a)$.

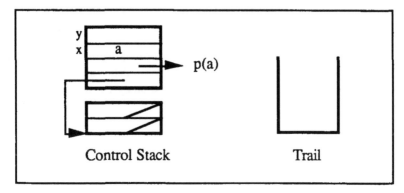

FIG. 4.6. An example of the use of the *trail* (b)

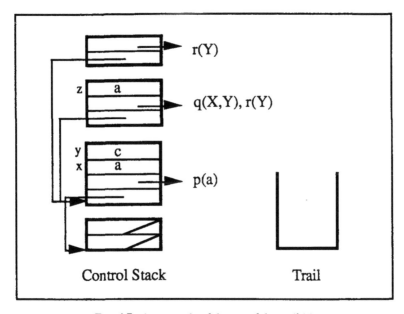

FIG. 4.7. An example of the use of the *trail* (c)

4.3.3. First Architecture for the Working Area

We get a first architecture for the working area in the form of two stacks updated only upon backtracking:

- the control stack, where the forward process allocates new activation blocks that will be popped upon backtracking; and

- the *trail*, where the postbound variables are saved in order to be restored to their initial free state upon backtracking.

But to continue with our study of better performing memory management schemes such as updating upon return from deterministic calls (chapter 6) or last-call optimization (chapter 7), we must first study the representation of structured terms in Prolog.

CHAPTER 5

REPRESENTATION OF TERMS

During the course of a proof, unification constructs new terms (instances) from those initially appearing in the knowledge base (models). Variables are bound to the Prolog terms, which are either simple (variables or constants) or structured (lists or functional terms). Whereas in the case of simple terms the binding occurs by direct association between the variable and the term, the situation is different for structured terms for which the binding can take place in two different forms: constructing a new instance starting from a model or accessing an already existing instance.

There are two different modes of representation for terms in Prolog: structure sharing and structure copying. If the purpose of both of these methods is to represent the new instances of the terms generated by the unification, they differ fundamentally by their principle and the binding mechanisms that they induce.

In structure sharing [BM72], [BM73], each instance of a structured term is represented in the form of a pair (skeleton, environment) where the environment provides the values of the variables appearing in the skeleton. For every binding between a variable and a structured term, the variable is associated with the pair (skeleton, environment) representing the term. The binding process works in an identical manner in both constructing and accessing.

Structure copying [Bru76], [Bru80], [Mel79], [Mel80], on the other hand, will generate every new instance of a structured term by copying the initial skeleton. This representation mode implies a different behavior for the binding algorithm. Indeed, in constructing, a copy of the skeleton is first made and then the variable is bound to this new term. On the other hand, in accessing, it is enough to associate the variable with the instance of the term already copied.

All existing Prolog systems can be characterized by the method they use for the representation of terms. We give here a preliminary classification of most Prolog systems (the list is not exhaustive):

1. structure sharing:

   - Prolog_1 [Rou75], [BM73];
   - Prolog_Dec10 [War77], [PPW78];
   - Prolog/P [CKC79], [BJM83], [Man84];
   - Prolog-II [Can82], [Can84], [Can86];
   - C-Prolog [Per84a];

- LisLog [BDL84];
- Prolog-Bordeaux [Bil84];
- Xilog [Xil84];

2. structure copying:

- Micro_Prolog [Cab81], [CC84];
- Quintus_Prolog [Qui86];
- SP-Prolog [SPP87];
- Turbo-Prolog [Man86];
- Delphia_Prolog [Man87];
- SB-Prolog [Deb88];
- AAIS-Prolog [AAI88];
- SISCstus Prolog [Car87];
- Arity Prolog [The88];
- BIM-Prolog [BIM86].

We will first introduce the method for binding a variable to a term. For this, we will distinguish between three major types of binding depending on whether the term is a constant, a variable, or a structured term. We will then study separately each of the two modes of representation of terms: structure sharing and structure copying.

5.1. Binding Mechanism

We first point out the two major requirements that the binding mechanism must satisfy and then distinguish three fundamental types of bindings depending on the nature of the terms involved: constant, variable, or structured term. We will then study each type of binding by mentioning the problems that can arise and by suggesting solutions to these problems.

5.1.1. Two Requirements

The binding mechanism must satisfy the following two requirements:

- it must be efficient in both creating and accessing in order not to penalize the unification; and
- it must be compatible with the stack structure as much as possible so as not to interfere with subsequent updates of the control stack.

Indeed, the binding of a variable to a term must be quick and, at the same time, must facilitate subsequent accesses to its value. In addition, no binding should generate a reference from the bottom to the top of the control stack. Any such reference will make future updates of the control stack impossible except upon backtracking.

5.1.2. Three Types of Bindings

We will distinguish three types of bindings between a free variable X and a term te, depending on the nature of te:

- type L1: te is a constant,
- type L2: te is a variable,
- type L3: te is a structured term.

For an L1 binding, the variable is linked to the appropriate constant. The cases L2 and especially L3 are more complex.

5.1.3. L2 Bindings

Prolog variables are used to transfer more or less complete information as unifications occur. The intervening role played by the unbound variables results in the creation of binding chains (see section 1.5.2). We will show how, on the one hand, we can respect the stack structure by choosing a sense for the binding and on the other hand, improve the access to the value of a variable. In this, the dereferencing process will help minimize the size of binding chains (see section 1.5.2).

5.1.3.1. Binding to a free variable

When binding two free variables, the reference is always created from the most recent variable to the older one. This choice respects the structure of the control stack by always creating references from the top to the bottom between the environments.

Let us look at the example in Figure 5.1. Only the state of the bindings resulting from the proof of the goal $p(b)$ is shown. Environments are numbered according to their creating time.

The decision to bind U (created at time 2) to Y (created at time 1) as well as Z (in $E3$) to U (in $E2$) allows us to respect the stack structure of the control area. However, binding chains appear, for example, on Z (length 1) or V (length 1).

5.1.3.2. Dereferencing

Dereferencing is the process of running through a chain of bindings in order to determine its extremity (see section 1.5.2). A systematic application of this

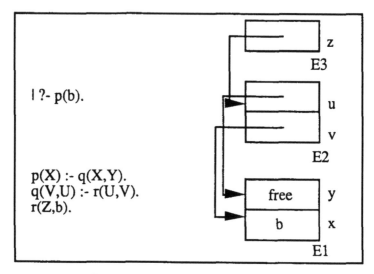

FIG. **5.1.** Binding between free variables

process minimizes the size of such chains, thus accelerating subsequent accesses to the values of the variables.

Consider once again the example in Figure 5.1, but this time apply the dereferencing (see Figure 5.2). The binding chain on Z disappears. In fact, when unifying Z and U, Z is bound to the dereferenced value of U, that is, Y. Similarly for V, which from now on directly designates the constant b.

Nevertheless, systematic application of dereferencing does not guarantee the complete absence of binding chains, as is shown in Figure 5.3.

We cannot avoid the chain of length 2 on V ($V = Y$, $Y = Z$, $Z = a$), and the chains of length 1 on U, Y, and X, all three bound to $Z = a$. In general, such chains arise from bindings between free variables to which only subsequently will unification assign nonvariable values.

5.1.4. L3 Bindings

When binding a free variable X of an environment Ei to a structured term te of an environment Ej, there are three cases to be considered depending on the respective creating times i and j of the two environments:

- type L3a: Ei is more recent than Ej or $Ei = Ej$, ($j \leq i$).

- type L3b: Ej is more recent than Ei, ($i < j$), but there are no variables V in te (te is a ground term).

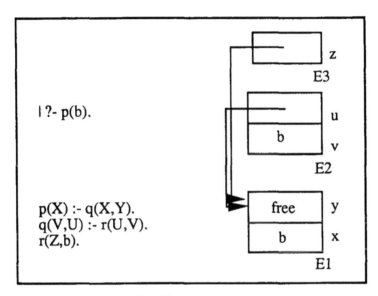

FIG. 5.2. Dereferencing

- type L3c: Ej is more recent than Ei, $(i < j)$ but there is at least one variable V in te.

Suppose we have to unify the two terms $t1 = f(X, Y, g(a, h(Z)), U)$ and $t2 = f(g(T, Q), L, g(T, R), h(c))$ in their respective environments $E1$ (allocated at time 1) and $E2$ (allocated at time 2) where all the variables are unbound.

The unification produces the following bindings: $X = g(T, Q)$ type L3c, $L = Y$ type L2, $T = a$ type L1, $R = h(Z)$ type L3a, $U = h(c)$ type L3.

This distinction aims at characterizing the behavior of a binding with respect to the creating times of the two environments involved. In the case L3a, the binding respects the creating times (X of Ei is bound to te in Ej where $j \leq i$), but this is reversed in the other two cases (L3b and L3c). Thus, the value of X will depend on the environment Ej, which is created subsequently, that is, is placed higher on the stack. This phenomenon generates bottom-up references on the control stack.

Finally, let us note that for a L3b binding the environment Ej is not involved in the representation of the term te since te is a ground term. There remains, a priori, one difficult case, the binding of type L3c.

But before proceeding with our presentation of this problem and its solutions (see chapter 6), we must describe the representation of structured terms in Prolog.

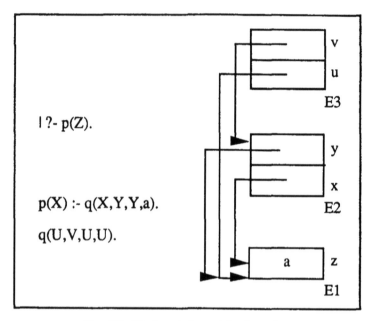

FIG. 5.3. Binding chains and dereferencing

5.1.5. Constructing and Accessing

An L3 binding can occur in constructing (creation of a new instance of the term) or in accessing (access to a subterm of an existing instance).

We will say that a structured term *te* is constructed [Mel82] if *te* becomes accessible, as the value of a variable X, for the first time. This phenomenon corresponds to the creation of a new instance of the term starting from a model appearing in the base of clauses.

When a term is matched against a structured term *te* that is already accessible, we will say that the components of *te* are accessed [Mel82].

Suppose we have to unify $f(X, h(Y), T)$ and $f(h(g(Z)), X, X)$. The unification produces the substitution $\{X = h(g(Z)), Y = g(Z), T = h(g(Z))\}$, where X is bound by constructing, and Y and T are bound by accessing. Indeed, the term $te = h(g(Z))$ becomes accessible for the first time when it is bound to the variable X and then Y and T access subterms of *te*, or *te* itself.

The distinction between these two forms of bindings is fundamental in the study of the representation of structured terms in Prolog. In fact, the first method of representation that we will discuss (structure sharing) induces an identical behavior in constructing and accessing, whereas the second one (structure copying) does not. For each of these two methods, we will first describe the principle and then show how Prolog uses it.

5.2. Structure Sharing

Structure sharing represents each instance of a structured term in the form of a molecule (skeleton, environment), where the environment describes the values for the variables occurring in the model. The technique of structure sharing was developed in 1972 by Boyer and Moore [BM72] and was introduced in a Prolog interpreter for the first time in 1973 at the University of Marseille by P. Roussel, G. Battani, and H. Meloni [BM73], [Rou75].

5.2.1. Principle of Structure Sharing

We will first describe the principle of structure sharing with the help of some examples and then we will make a preliminary evaluation of the behavior it induces regarding the access and creation of terms.

5.2.1.1. The Principle

In structure sharing [BM72], an instance of a term is represented by a pair (skeleton, environment) where the skeleton is the model of the term and the environment provides the values of the variables appearing in the skeleton.

Examples:
1. Consider the skeletons $s1 = f(g(X), Y)$ and $s2 = h(X)$ and the environment $E = \{(X, b); (Y, (s2, E))\}$. Then $t = (s1, E)$ represents the term $t = f(g(b), h(b))$.
2. Given $s1 = f(g(X), Y)$ and $s2 = h(X, T)$ and the environments $E1 = \{(X, b); (Y, (s2, E2))\}$ and $E2 = \{(X, a); (T, (X, E2))\}$. Then $t = (s1, E1)$ represents the term $t = f(g(b), h(a, a))$.
3. Given $s1 = f(g(X), Y, Z)$ and $s2 = h(U, V)$ and the environments $E1 = \{(X, (s2, E2)); (Y, (s2, E3)); (Z, (s2, E2))\}$, $E2 = \{(U, a); (V, b)\}$, $E3 = \{(U, c); (V, d)\}$.
   Then $t = (s1, E1)$ represents the term $f(g(h(a, b)), h(c, d), h(a, b))$.

When the skeleton is reduced to a constant, it is not necessary to specify the environment. The complete instance of a term $te = (sq, E)$ is obtained by applying to the skeleton sq the substitution represented by the environment E.

The distinctive feature of this form of representation is that different instances of terms can share same models, or structures (hence the name). Two instances of terms with the same model can differ only by their environment. In the third example, the skeleton $s2$ is shared by the three variables X, Y, and Z. Depending on the environment in which it is taken, it gives rise to the term $h(a, b)$ or to the term $h(c, d)$.

5.2.1.2. Preliminary Evaluation

Structure sharing manages all the L3 bindings uniformly, whether they occur in constructing or accessing (see section 5.1.5). In fact, every binding of a variable to an instance of a structured term is done by assigning a molecule (skeleton, environment). The space needed for the representation of a term is, in the worst case, proportional to the number of distinct variables appearing in the skeletons rather than proportional to the sum of the sizes of these skeletons.

  Therefore structure sharing allows for quick and economical creation of new terms. On the other hand, access to a complete instance of a term is very often slowed down by the successive intermediate references that must be traversed, as is shown in first example of the previous subsection: to obtain the instance of $t = f(g(X), Y)$ in E, we must traverse $(X, E) = b$, followed by $(Y, E) = (h(X), E)$, and then again $(X, E) = b$.

  The second version of the unification algorithm given in section 1.5.2 and also re-used in section 2.4.6 uses a particular implementation of structure sharing. Indeed, the instances of terms are created directly from the models without any copying done. But, unlike in the principle discussed above, the environments are not separate but grouped together in one. In this case, it is no longer necessary to specify the environment as the second part of the molecule since we are always referring to the same one. A direct binding to the model is sufficient.

5.2.2. Using Structure Sharing in Prolog

Each time a goal is called, the duplication of the clause candidate for the unification allocates a new environment on the control stack (see section 4.2.1). This environment is a vector of size n (the number of distinct variables occurring in the clause) where each elementary zone is a molecule di that will receive the value of the corresponding variable vi in the form of a pair (skeleton, environment). All variables are initially free. This state will be symbolized by the "binding" to a particular atom, the atom $free$.

  Since the variables are numbered from 0 to $n - 1$, access to the ith variable in the environment e is achieved by using i as an index into e. Skeletons of terms are included in the coding area of the clauses. Environments are allocated on the control stack.

  Figure 5.4 shows the bindings resulting from the proof of the goal $p(a)$ (only the environments are shown). At each duplication, a new environment is allocated. Here, the numbering of the variables in an environment is the same as their order of appearance in the clause, that is, $V0 = X$, then $V1 = Y$ for $E1$ and $V0 = Z$, then $V1 = T$ for $E2$.

  To bind a variable amounts to assigning a pair of references to its molecule,

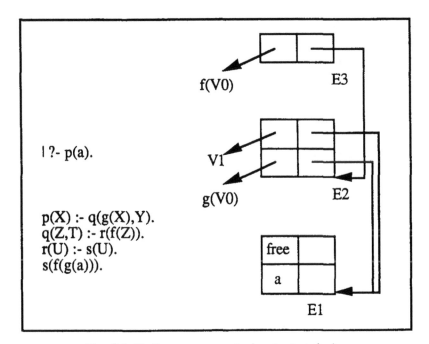

FIG. 5.4. Bindings management using structure sharing

one reference to the skeleton in the coding area of the clauses and another to the associated environment on the control stack. Here the special atom *free* indicates the presence of a free variable (this is the case with $V1 = Y$ in $E1$).

The creation of new instances of terms is therefore based directly on the term models.

5.3. Structure Copying

Structure copying constructs new instances of terms by copying their models taking into account the values of the variables appearing in them. This method of representation of terms has been introduced by M. Bruynooghe [Bru76], [Bru80], and C. Mellish [Mel79], [Mel80].

Structure copying manages the bindings differently depending on whether structured terms are constructed or accessed (see section 5.1.5). In the first case, each structured term is first copied and then the variable is bound to this copy. In the second case, the binding is direct without prior copying. Using such a method of representation of terms requires the introduction of a new memory area, the copy stack (or heap), into the working area of the Prolog system.

Our approach will be the same as in the case of structure sharing: we will first present the principle of structure copying and then we will study the way in which Prolog utilizes it.

5.3.1. Principle of Structure Copying

The basic principle of this mode of representation [Bru76], [Bru82a], [Mel79], [Mel82] is, for each binding by constructing, to make a copy of the model by substituting the variables appearing in it by their values. Thus, unlike structure sharing (see section 5.2.1), bindings by constructing are managed differently than bindings by accessing.

5.3.1.1. Copying Algorithm

Let us first define the binding mechanism [Bru82a] between a variable x and a term te:

- if te is a constant or a variable, x is directly bound to te;
- if te is an instance of a structured term (i.e., a substituted copy of a term model), x is also directly bound to te (accessing);
- if te is a term model, then x is bound to a copy of te (constructing).

The copying of the term model te in an environment e is done as follows [Bru82a]:

$copy\,(te, e)$
    begin
        three cases depending on nature of te:
            1. te is a constant then te;
            2. te is a variable
                let tx be the dereferenced value of te;
                if tx is a free variable,
                    then generate a new free variable x;
                        bind tx to this newly created variable x;
                  else tx.
                endif;
            3. te is a structured term
                let f be the functional symbol for te and n its arity;
                re-copy f/n;
                for $i := 1$ to n do $copy(i^{th}\ argument\ of\ te, e)$;
    end;

In the second case, if tx is not a free variable, we are sure that tx is either a constant or an instance of a structured term (therefore already copied). In the third case, the n arguments of the term are recursively copied.

Examples:

1. Suppose we have to unify the two terms $t1 = f(g(X), Y)$ and $t2 = Z$ in the environment $E = \{(X, c); (Y, b)\}$. Z is bound to a copy of $t1$ where the variables X and Y have been substituted by their value in E, that is, $Z = f(g(c), b)$.

2. Suppose we have to unify $t1 = f(g(X), Y)$ and $t2 = f(Z, U)$ in E empty. Z is then bound to a copy of $g(X)$ in E. Since X is free, we generate a new variable $X1$ and X is bound to $X1$. E becomes $\{(Y, U); (Z, g(X1)); (X, X1)\}$.

3. Suppose we have to unify, in an empty environment E, $t1 = f(X, g(U, W), Y)$ and $t2 = f(T, Z, g(V, g(V, Z)))$. Unification between $t1$ and $t2$ produces the new environment: $\{(Y, g(V1, g(V2, g(U1, W1)))); (V1, V2); (V, V1); (Z, g(U1, W1)); (W, W1); (U, U1); (X, T)\}$

In fact, X is first of all bound to T, then Z is bound to a copy of $g(U, W)$, and finally Y is bound to a copy of $g(V, g(V, Z))$. For this, we create $V1$ and we bind V to $V1$. But $V1$ is also a free variable, so we generate $V2$, and $V1$ is bound to $V2$. Finally, since the value of Z is already a copy, it is replaced directly.

As example 3 shows, structure copying may generate several instances of a same free variable. All these instances must be correctly linked, so that they always designate the same object.

5.3.1.2. A Preliminary Evaluation

Structure copying implies a different behavior of the binding algorithm depending on whether it is accessing or constructing. Because of this, structure copying facilitates accesses to instances of structured terms. On the other hand, the creation of new terms from models requires copying the models.

The space needed for the representation of a term becomes at worst proportional to the sum of the sizes of its subterms. In addition, and unlike the case of structure sharing (see section 5.2.1), the same variable may be allocated several locations. For this to happen, it is enough that the variable be in a free state at the time of the term model, in which it occurs, is copied.

5.3.2. Implementation

Let us implement in Lisp the unification algorithm using structure copying with the same conventions and coding as in section 1.5.

Structured terms are represented in the form of lists. A variable is a Lisp atom beginning with '_'. Environments are coded using A-lists. The following definitions remain unchanged:

```
(defconstant marque #\_)
(defun var? (x)
   (and (symbolp x) (char= (char (string x) 0) marque)))
(defun value (x s) (cdr (assoc x s)))
(defun bound? (x s) (assoc x s))
(defun add (x val s) (acons x val s))
(defun val (te s) (if (var? te) (ult te s) te))
(defun ult (v s)
   (if (bound? v s) (val (value v s) s) v))
```

When binding a variable x to a term sq, we must henceforth distinguish four cases depending on the nature of sq: variable (L2 binding), constant (L1 binding), instance of a term (L3 binding by accessing) or model for a term (L3 binding by constructing).

```
(defun bind (x sq s)
   (if (or (var? sq) (atom sq) (copy? sq))
       (add x sq s)
       (multiple-value-bind (te e)
          (copy sq s)
          (add x te e))))
(defmacro copy? (x)
   '(eql (car ,x) #\*))
(defmacro tcons (x y)
   '(cons #\* (cons ,x ,y)))
```

The first three cases are treated identically. On the other hand, if sq is a term model, we copy it (te) and then bind the variable x to the new instance te. To distinguish models for terms from instances, we will tag the lists molecules representing the latter.

Copying a model involves providing a duplicate where all the variables are substituted using the algorithm described in section 5.3.1. (copy x s) generates recursively a copy of x in the environment s while tagging the lists (tcons). For a free variable a, we generate a new unbound variable b and then bind a to b. If the variable is bound to a constant or an instance of a structured term, we consider its value.

```
(defun copy (x s)
   (cond
      ((var? x)
          (let ((a (ult x s)))
             (if (var? a)
                 (let ((b (genvar a)))
                    (values b (add a b s)))
```

```
                    (values a s))))
           ((atom x) (values x s))
           (t (multiple-value-bind (t1 e1) (copy (car x) s)
                (multiple-value-bind (t2 e2) (copy (cdr x) e1)
                  (values (tcons t1 t2) e2))))))))

(defun genvar (x)
   (gentemp (concatenate 'string (string x) "-")))
```

The copy function yields two results, the copy of the term and the new environment. These two values are handled by the *multiple-value* mechanism of Common-Lisp.

For a structured term, first the car is copied into the environment s. Then, cdr is copied into the new environment e1. The final result is the term obtained from t1 and t2 and the environment e2. Finally, new free variables generated from existing ones keep the same name but are numbered differently (see genvar definition).

```
(defun unif (x y s)
   (let ((x1 (val x s)) (y1 (val y s)))
      (cond
           ((eql x1 y1) s)
           ((var? x1) (bind x1 y1 s))
           ((var? y1) (bind y1 x1 s))
           ((or (atom x1) (atom y1)) 'fail)
           (t (let ((news (unif (mycar x1) (mycar y1) s)))
                (if (eq news 'fail)
                    news
                    (unif (mycdr x1) (mycdr y1) news)))))))
(defmacro mycar (x)
   '(if (copy? ,x) (cadr ,x) (car ,x)))
(defmacro mycdr (x)
   '(if (copy? ,x) (cddr ,x) (cdr ,x)))
```

The unification is unchanged (see section 1.5.2) except that:

- binding a variable to a term is no longer necessarily done by add, and
- when two structured terms are unified, it is necessary to ignore the tagging of lists molecules representing instances of terms (mycar, mycdr).

Finally, let us pretty-print the unifier obtained in this way:

```
(defun test (x y)
   (let ((teta (unif x y ())))
      (if (eq teta 'fail) 'fail
          (mapc
            #'(lambda (x)
                (format t "~A = ~A~%" (car x) (ext (cdr x) teta)))
            teta))))
```

```
(defun ext (v s)
   (if (atom v)
       v
       (cons (ext (cadr v) s) (ext (cddr v) s)))))
```

Let us test the three examples described in section 5.3.1:

```
> (test '(t _X _Y (f (g _X) _Y)) '(t c b _Z))
  _Z = (f (g c) b)
  _Y = b
  _X = c
> (test '(f (g _X) _Y) '(f _Z _U))
  _Y = _U
  _Z = (g _X-1)
  _X = _X-1
> (test '(f _X (g _U _W) _Y) '(f _T _Z (g _V (g _V _Z))))
  _Y = (g _V-4 (g _V-4-5 (g _U-2 _W-3)))
  _V-4 = _V-4-5
  _V = _V-4
  _Z = (g _U-2 _W-3)
  _W = _W-3
  _U = _U-2
  _X = _T
```

5.3.3. Using Structure Copying in Prolog

Using structure copying in Prolog needs to introduce into the working area of
the Prolog system a new memory area usually called copy stack or heap. We
will first describe this new memory organization and the copying algorithm,
followed by some examples.

5.3.3.1. Heap or Copy Stack

Each time a goal is called, the duplication of the clause candidate for the unification allocates a new environment on the control stack. This environment is
a vector of size n (number of distinct variables occurring in the clause) where
each elementary zone is a simple location where the value of the variable will
be placed. The initial unbound state of every variable is obtained by "binding"
every variable to itself.

But, being of fixed size, the environment cannot hold all the copies of the
terms eventually created. Each copy is made in a special memory area called
copy stack or heap. This double naming comes from the fact that this zone is
updated like a stack but in reality has a heap structure.

Indeed, its management is similar to that of a stack. Each time a copy of a
term t is made, as many memory locations as required are allocated to the top of
the stack. Upon backtracking, the copy stack is updated by simply restoring the
register designating its top. But, the copying algorithm will generate top-down

and bottom-up references in this memory area which, therefore, really has a
heap structure.

5.3.3.2. Copying Algorithm

Consider again the copying algorithm given in section 5.3.1. The replacement of
a variable X appearing in the term t (to be copied according to the environment
E) is done as follows:

begin
    let tx be the dereferenced value of X in E;
    if tx is a free variable,
        then allocate a new unbound variable at the top of the copy stack;
            link tx to this location;
        else
            allocate a new location at the top of the copy stack;
            link this location to tx.
    endif;
end;

This algorithm generates bottom-up and top-down references on the copy
stack. Indeed, in copying a term te containing at least one variable X bound to
an instance of a structured term tx, a reference is created from the top toward
the bottom. On the other hand, in binding a free variable X, held on the copy
stack, to a model for a term te (thus requiring a copy tx to be made at the top
of the stack), a reference is created from the bottom toward the top.

The space needed for coding the bindings between variables and values is
therefore divided into two parts: a vector of size n allocated on the control stack
used as an entry table for finding the value of a variable, and locations scattered
around in the copy stack.

5.3.3.3. Examples

Let us consider again the example in Figure 5.4, but now using a structure copying approach (see Figure 5.5). Each time a goal is called, the clause candidate
for the unification is duplicated by allocating a vector of size n on the control
stack (the variables are still numbered according to their order of appearance).

When $q(g(X), Y)$ is called, Z (the first variable in $E2$) is bound to a copy
of the model $g(X)$ in the environment $E1$ (g is of arity 1). Then U, the only
variable in $E3$, is bound to a copy of the model $f(Z)$ in the environment $E2$ (f
is of arity 1). A reference to the value of Z on the copy stack is created. Finally,
the unbound state of the variable Y (second variable in $E1$) is indicated by the
fact that it is "bound" to itself.

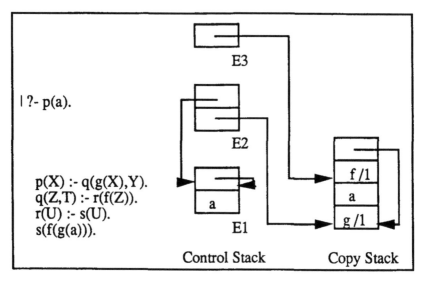

FIG. 5.5. Binding management using structure copying

Let us look at another example. Figure 5.6 shows the state of the bindings after the proof of the goal $q(X, f(Y))$.

V (second variable in $E2$) is bound to a copy of the model $f(Y)$ in $E1$. Since Y (second variable in $E1$) is free, a new unbound variable is allocated at the top of the copy stack. Then Y is linked to this memory location.

Figure 5.7 describes the state of bindings just after the call to $r(g(h(Y)))$. The variable Z is bound to a copy of the model $g(h(Y))$ in $E1$. The two locations designating Y on the copy stack are linked together (see section 5.3.1).

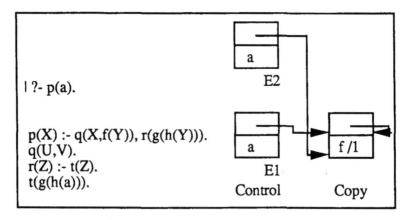

FIG. 5.6. Example of structure copying (a)

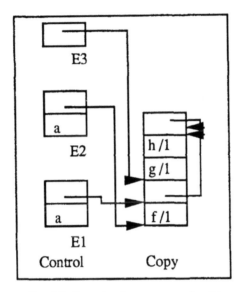

FIG. 5.7. Example of structure copying (b)

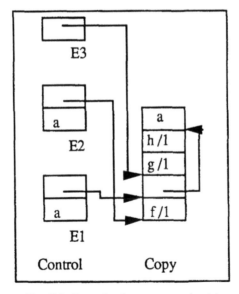

FIG. 5.8. Example of structure copying (c)

The proof then ends (see Figure 5.8) binding Y to constant a when the goal $t(Z)$ is called.

The use of structure copying thus implies an architecture of the working area divided into three stacks (control + copy + *trail*) while structure sharing seems to require only two stacks (control + *trail*).

But before continuing on with the comparison between these two modes of representation for terms (see section 6.4), we must study how memory space can be reclaimed upon return from deterministic calls.

CHAPTER 6

DETERMINISTIC CALL RETURNS

The updating of the control stack upon return from deterministic calls forces the introduction of a third stack in Prolog's working area, independent of the terms representation's choice, whether by structure sharing or structure copying.

Indeed, since L3c bindings do not respect creation times (see section 5.1.4), a variable could have a life span greater than that of the clause where it occurs. A premature updating could then cause an irrecoverable loss of information. Such bindings should therefore be allocated in a special memory area, outside the control stack.

In the case of structure sharing, the proposed solution in Prolog-Dec10 [War77] consists of statically identifying the variables capable of generating such bindings. They are allocated an environment outside the control stack, in a special area called the *global stack*.

In the case of structure copying, the problem is automatically solved since every binding to a model of a structured term is achieved by copying the latter in the memory area reserved for that purpose (see section 5.3.3).

In both cases, we find a three-stack architecture that forms the working area: a control stack (*local stack*), a *global stack* or a copy stack, and a *trail*. The first one is updated upon return from deterministic calls and upon backtracking, whereas the other two stacks are updated only upon backtracking.

We will first present the problem that arises in L3c bindings. We will then describe the Prolog-Dec10 solution in the context of structure sharing. Afterward, we will propose an equivalent approach in the case of structure copying. Finally, we will conclude by comparing both proposals.

6.1. The Problem

Upon return from a deterministic call, and in order to recover the space allocated on the control stack for its execution, one should ensure that the updating does not result in an irreversible loss of information. It is particularly the case if all references are made from the top of the stack to its bottom. The references stated by the continuation and rerun parts satisfy this need (see section 4.2). This question then remains unanswered only for the environments.

In the case of a structure sharing, L3c bindings (see section 5.1.4) will generate bottom-up references between the control stack environments, thus forbidding all updatings outside backtracking.

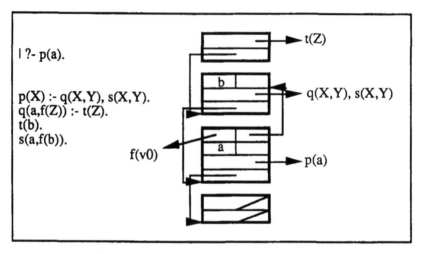

FIG. 6.1. An example of bottom-up references (a)

After describing the problem using an example, we will present a first solution, which consists of splitting environments from the rerun and continuation parts that supervise the control.

6.1.1. Example

Let us study an example where an L3c binding builds a bottom-up reference, thus forbidding any premature updating.

Figure 6.1 indicates the control stack status after calling the goal $t(Z)$. Four deterministic blocks have been allocated (see section 4.2.3). The first block refers to the query. The second one, due to the calling of $p(a)$, has an environment of size 2 representing X and Y, where X is bound to the constant a. When calling the goal $q(X, Y)$, Y is bound to the structured term $f(Z)$, through the mapping of the molecule that represents it (see section 5.2.2). The activation of the goal $t(Z)$ then binds Z to the constant b.

The call to $t(Z)$ being deterministic and completed, the control stack is updated upon return (see Figure 6.2).

Similarly, the call to $q(X, Y)$ being deterministic and completed, we should be able to recover its activation block from the control stack (see Figure 6.2). However, the L3c binding between Y and $f(Z)$ forbids such an updating. Thus, the creation of L3c bindings prevents all updating upon return from deterministic calls.

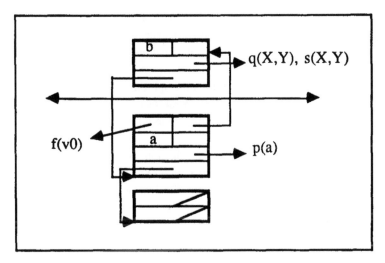

FIG. 6.2. An example of bottom-up references (b)

6.1.2. First Solution

In order to solve this problem in the context of structure sharing, one solution [CKC79] consists in splitting each activation block according to the environment and control parts. The working area then has a three-stack architecture:

- a control stack, holding the continuation and rerun parts, updated upon backtracking and upon return from deterministic calls,

- a substitutions stack, holding the environments and only updated upon backtracking,

- a *trail*, updated upon backtracking.

The concept of separating the environment and control parts was used further by M. Van Caneghem [Can84], [Can86] during the implementation of Prolog-II . But this choice is also mainly due to the implementation of the delayed mechanism associated to the predicates $dif/2$ and $freeze/2$. This will be pursued further in chapter 10. Such a management of the substitutions stack keeps numerous bindings, especially all those not of L3c type that have become useless with respect to the rest of the proof.

6.2. Solution for Structure Sharing

The fact that we are able to determine statically the variables capable of generating L3c bindings brought D. H. Warren, while implementing Prolog-Dec10

[War77], to manage them in a special way by allocating them a distinct environment outside the control stack. Since all danger of bottom-up referencing has been avoided, the control stack can be updated upon return from deterministic calls.

The consequence of this solution is the splitting of each environment into two parts. The first part is held by the control stack. The second part, associated with the variables that might generate L3c bindings, is stored in a special area labeled *global stack*.

After presenting the variables classification proposed by D. H. Warren [War77], we will describe the new architecture of the working area, together with its respective memory management. Finally, we will conclude by recalling the example of Figure 6.1 and by demonstrating that the updating is now valid.

6.2.1. Variables Classification

D. H. Warren proposes classifying the variables occurring in a Prolog clause in four main categories: global, void, temporary, and local. This classification is based upon the time interval during which the information held by the variables is useful. We will provide some examples after describing the four variables categories.

6.2.1.1. Four Categories of Variables

Let cl be a Prolog clause. We can then classify all variables occurring in cl according to four categories [War77]:

- X is a global variable iff X occurs at least once in a structured term;
- Among the variables left (they all appear at the first level in cl), we can distinguish three types:

  - The void variables: X is a void variable iff X has only one occurrence in cl;
  - The temporary variables: X is a temporary variable, iff X appears more than once in cl, with no occurrence in the body;
  - The local variables: X is a local variable, iff X appears more than once in cl, with at least one occurrence in the body.

The global variables are thus the only ones capable of generating L3c bindings, since they are the ones that appear in the structured terms (see section 5.1.4).

6.2.1.2. Examples

In the clause

```
p(X,Y,f(Z),Y) :- q(Z,T,U), r(X,U)
```

Z is then global, X and U local, Y temporary, and T void.

Let us consider the quicksort with accumulator version as defined by [War77] and [SS87]:

```
partition([],N,[],[]).
partition([X|L],Y,[X|L1],L2) :-
    le(X,Y), !,
    partition(L,Y,L1,L2).
partition([X|L],Y,L1,[X|L2] :-
    partition(L,Y,L1,L2).
qsort([],R,R).
qsort([X|L],R,R0) :-
    partition(L,X,L1,L2),
    qsort(L2,R1,R0),
    qsort(L1,R,[X|R1]).
```

In the definition of partition/4, N is a void variable for the fact. Y and L2 are local in the first rule, while X, L, and L1 are global. In the second rule, Y and L1 are local, whereas X, L, and L2 are global. In the definition of qsort/3, R is temporary for the fact. R, L1, L2, and R0 are local in the rule, whereas X, L, and R1 are global.

The number of variables within each category varies according to the nature of the Prolog programs. In general, the void variables and the temporary variables are quite rare, and most of the time they are related to facts. The ratio between local and global variables depends upon the problem dealt with and upon the use of structured terms.

This classification has been achieved by observing that each category of variables has a very specific life span. Indeed, a void variable will satisfy any unification and will never be used afterward. We can then consider its life span to be null. A temporary variable will only be useful during unification between a goal and the head of the corresponding clause. Therefore, its life span is that of the unification.

A local variable will have a life span identical to that of the rule to which it belongs. It might be used all through the execution of the body of the rule, but it will become unreachable once the proof ends.

Finally, a global variable—since it appears in a structured term—is capable of generating L3c bindings and is thus susceptible to having a life span longer than that of the clause where it occurs.

6.2.2. Architecture and Memory Management

The solution proposed by D. H. Warren in Prolog-Dec10 [War77] leads to a three-stack architecture of the working area:

- a control stack, or *local stack*, updated upon return from deterministic calls and upon backtracking;
- a *global stack* updated upon backtracking; and
- a *trail* updated upon backtracking.

First, we will present the new concept of environment introduced by the distinction between local and global variables. We will then describe the activation blocks. Finally, we will conclude by adapting the binding mechanism (see section 5.1) and the concept of postbound variables (see section 4.3.1) to this new situation.

6.2.3. Environments

Each environment is now split into a local and a global part. Indeed, let cl be a clause containing nl local variables and ng global variables. The local variables are numbered from 0 to $nl - 1$ and the global ones from 0 to $ng - 1$. The clause duplication this time produces (see section 4.2.1) the 3-uple $(cl, Elocal, Eglobal)$, where $Elocal$ and $Eglobal$ are two vectors of length nl (ng) respectively and are to receive the values for the nl (ng) local (global) variables.

The duplication of the rule cl:

```
qsort([X|L],R,R0):- partition(L,X,L1,L2),
    qsort(L2,R1,R0),
    qsort(L1,R,[X|R1]).
```

produces the 3-uple (cl, El, Eg) where cl is the skeleton of the rule, El and Eg two new environments of size 4 and 3 respectively—since R, R0, L1, and L2 are local variables, whereas X, L, and R1 are global ones.

The local parts of the environments are held by the activation block of the control stack, whereas the global parts are allocated in the *global stack*. We have voluntarily omitted a discussion of the management of void and temporary variables. Let us see now how these are handled by the Prolog-Dec10 compiler [War77].

A void variable has no memory allocation. Temporary variables are managed as "fake" local variables from which we regain—after the unification process—the space allocated in the control stack.

Indeed, let there be a call for a goal b and let cl be the candidate clause for unification. Let nl, ng, and nt be the number of local, global, and temporary variables respectively. The duplication of cl produces the 3-uple (cl, El, Eg) where El is a vector of length $nl + nt$ and Eg a vector of length ng. Practically, the local variables are numbered 0 to $nl - 1$ and the temporary ones from nl to $nl + nt - 1$. If the unification between b and the head of cl succeeds, the space allocated for the temporary variables is freed, and processing continues.

This update is possible because, as we shall see, the local part of the environment is allocated at the top of the activation block. The block is freed by updating the top of the stack register, at the position occupied by the first temporary variable (i.e., nl) and not just after the position of the last temporary variable (i.e., at $nl + nt$).

6.2.4. Activation Blocks

Let us consider again the composition of the activation blocks described in section 4.2.2, and let us adapt it to this new situation.

6.2.4.1. Allocation of an Activation Block

Given the call to the goal b_i with related siblings $b_{i+1}, \dots, b_n$, let cl be the clause whose head satisfies the unification; hence:

- the continuation part remains the same (see section 4.2.3):
  - CP designates the sequence of goals $b_i, b_{i+1}, \dots, b_n$,
  - CL indicates the father block of this sequence of goals;
- the environment part is now as follows:
  - on the *local stack*:
    * a field G, which indicates the beginning of the global part of the environment,
    * a field E, made up of a length nl vector (number of local variables occurring in the clause cl);
  - on the *global stack*:
    * a vector of length ng (number of global variables occurring in the clause cl).

For a choice block, the rerun part remains the same (see section 4.2.4):

- the field BL indicates the previous choice block on the *local stack*,
- the field BP designates the remaining choices as a set of clauses still to be considered, and
- the field TR describes the top of the *trail* before the allocation of the choice block.

The continuation and rerun parts remain unmodified. While for the continuation part the result is immediate, it is not for the rerun part.

6.2.4.2. Backtracking

The three stacks should be updated upon backtracking. While for the *local stack* the BL register provides the required value (last choice block) and the TR field of the block BL does the same for the *trail*, no information is explicitly available for the updating of the *global stack*.

Indeed, the required value is found in field G of block BL, since this field indicates the beginning of the global environment of the block, that is, the top of the *global stack* at the time of the creation of the point of choice. Usually [War77], this information is stored in a particular register BG, even though it is accessible through BL. The last choice point is then described by the couple (BL, BG).

In case of successive failures, the different backtrackings are performed by updating the three stacks (the *local stack* by using BL, the *global stack* by using BG, and the *trail* by using the TR field of block BL), together with the BL (using the BL field of block BL) and the BG (using the G field of block BL) registers.

Finally, under this format, the rerun part of a choice block is not capable by itself of providing all the necessary information to restart the forward process. It first borrows the goal to be proved (field CP) from the continuation part (see section 4.2.4) together with the environment within which it should be considered (accessible through the CL field). Furthermore, it uses the G field of the environment part in order to update the *global stack*.

6.2.5. Binding Mechanism

The difference between local and global variables insures no bottom-up referencing on the local stack. In order to achieve the updating of the latter upon return from deterministic calls, we should also guarantee that there exist no references from the *global stack* to the *local stack*. The only case susceptible to producing such a binding is that of the unification between two unbound variables—thus of type L2 (see section 5.1.3)—where one is local and the other global.

In this case, since all L2 bindings preserve creation times (see section 5.1.3), it is sufficient to consider a global variable to be always older than a local one. This problem is usually solved [War77], [War80] by considering the working space an adjacent memory area, divided into three sections representing the three stacks (Figure 6.3).

Note that in the case of an adjacent memory representation, the creation time of a variable can be deduced from the address of the memory location

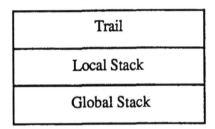

FIG. 6.3. Working area for structure sharing

representing it. Consequently, the *global stack* shall then be located under the *local stack*. As such, a global variable will always be older than a local one. The position of the *trail* is not important, but it is usually found at the top of this area.

6.2.6. Postbindings

The notion of postbinding (see section 4.3.1) should be extended to the global variables. A variable should be saved on the *trail* if it is prior to the last choice point (see section 4.3.1). With respect to a local variable, this corresponds to the case where it is located under the last choice point (register BL), and, for a global one, to the case where it is located under the global environment of the last choice point (register BG).

6.2.7. Example

Let us review the example described in Figure 6.1. This time, Z will be a global variable, and updating can then be performed. After the call to $t(Z)$, four deterministic blocks have been allocated (see Figure 6.4). The first two blocks have an empty global environment (not shown on the figure). The third deterministic block identifies with the duplication of the clause $q(a, f(Z)) : - t(Z)$ and has only a global environment, which is held by the *global stack*. Y is bound to the model $f(Z)$, which will be considered this time in the global environment.

The stack is popped upon return from $t(Z)$. Similarly, since the call to $q(X, Y)$ is deterministic and completed, its activation block is gained back. This leads to the status of Figure 6.5.

Since Z has been given a global variable status, the problem is solved. Indeed Z has life span superior to that of the clause in which it occurs.

Let us end the proof successfully (see Figure 6.6). Since the $s(X, Y)$ and $p(a)$ calls are deterministic, the two blocks can be freed. The global environment

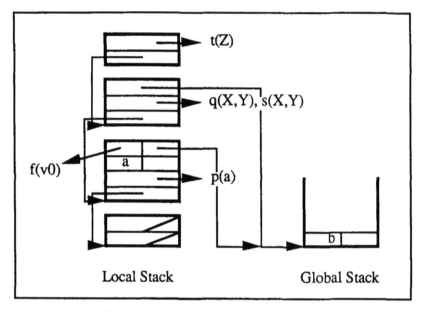

FIG. 6.4. Updating using structure sharing (a)

associated with Z becomes useless but is kept on the *global stack*, which itself is only updated upon backtracking.

6.3. Solution for Structure Copying

The problem described in section 6.1 is inherent to structure sharing and does not exist in a term representation that uses structure copying. Indeed, since any binding to a model of structured term is achieved by copying in a special memory area (see section 5.3.3), no bottom-up references can be generated between the environments on the control stack. The latter can thus be easily updated upon return from deterministic calls.

6.3.1. Architecture and Memory Management

The working area is divided into three stacks (see section 5.3.3):

- a control stack, or *local stack*, updated upon return from deterministic calls and upon backtracking;

- a copy stack, or *global stack*, updated upon backtracking; and

- a *trail* updated upon backtracking.

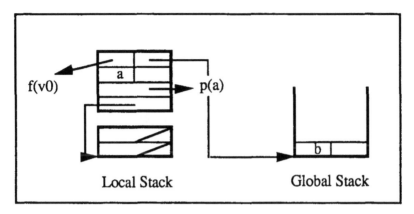

FIG. 6.5. Updating using structure sharing (b)

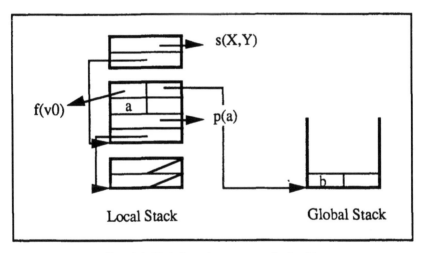

FIG. 6.6. Updating using structure sharing (c)

Primarily, we will recall the environments description for structure copying. We will then describe the composition of the activation blocks. Finally, we will conclude by adapting the binding mechanism (see section 5.1) and the concept of postbinding (see section 4.3).

6.3.2. Environments

On each duplication of a clause cl holding n variables, a vector of size n is allocated on the control stack (see section 5.3). This vector is made up of n elementary locations intended to receive the values of the n variables. Thus, as bindings are built, the necessary copies will be allocated on the *global stack*.

A complete environment is divided into two parts:

- a local part that is somehow used as a look-up table to access the values of the variables; this vector is allocated upon duplication of the clause; and

- a global part held by the copy stack; this part is distributed on the *global stack*, since it is allocated as the necessary copies (resulting from bindings to structured term models) are needed.

6.3.3. Activation Blocks

Let us reconsider the form of the activation blocks described in section 4.2.2, while adapting it to this new situation. Given the call of goal b_i with related siblings $b_{i+1}, ... , b_n$, let cl be the clause whose head satisfied the unification. Hence:

- the continuation part is unchanged (see section 4.2.3), and

- the environment part is composed of a vector of size n (number of variables occurring in cl).

For a choice block, we have to add a supplementary field to the rerun part (see section 4.2.4). Indeed, the three stacks must be updated upon backtracking. If the value for the *local stack* is provided by the BL register (last choice block), and the value for the *trail* is provided by the TR field of the BL block, no explicit information is kept, however, for updating the *global stack*.

Unlike structure sharing (see section 6.2.4), we cannot find the right value by accessing the environment. We should then make this information appear explicitly. Consequently, a fourth field (BG) is added to the existing three in each rerun part (see section 4.2.4):

- the BL field (previous choice block on the *local stack*),

- the BP field (remaining choices in the form of the set of clauses still to be considered),

- the TR field (top of the *trail*), and

- the BG field (describing the current top of the copy stack before the allocation of the choice point).

The last choice point is always characterized (see section 6.2.4) by the couple (BL, BG), where the BG register designates the top of the copy stack associated with BL.

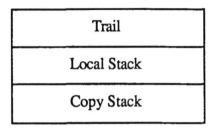

FIG. 6.7. Working area for structure copying

6.3.4. Binding Mechanism

The copying algorithm (see section 5.3.3) guarantees the absence of bottom-up references on the *local stack*. In order to perform the updating of the latter upon return from deterministic calls, we should also guarantee that there exist no references from the copy stack toward the *local stack*. The only case where such a binding could occur is that of unifying two free variables (see section 5.1.3), where one variable is held by the *local stack*, and the other by the copy stack.

Since any L2 binding is made while respecting the creation time (see section 5.1.3), it would be enough to consider a variable that appears in the copy stack to be always older than a variable held by the *local stack*.

This problem is solved as for structure sharing [War83] by considering the working area as a contiguous memory zone divided into three stacks, since the copy stack is located below the *local stack* (see Figure 6.7). A variable on the copy stack will then always be older than a local one.

6.3.5. Postbinding

Let us now adapt the concept of postbinding (see section 4.3.1) to the variables located in the copy stack. A variable must be saved on the *trail* if it is older than the last choice point. This is the case for a variable held by the *local stack* that is located below the last choice block (BL register), or for a variable prior to BG held on the copy stack.

6.3.6. Example

Let us consider again the example described in Figure 6.1, previously developed for structure sharing (see section 6.2.7).

The binding of Y triggers the copy of the term $f(Z)$. A new location representing a free variable is allocated on the copy stack, and Z designates that location (see Figure 6.8). Thereafter, during the proof of $t(Z)$, Z is bound to the constant b.

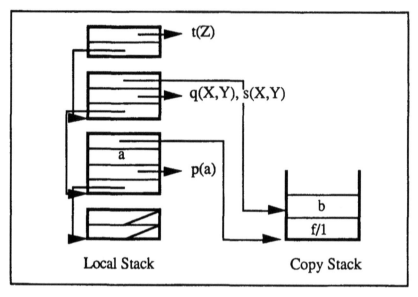

FIG. 6.8. Updating using structure copying (a)

The *local stack* is updated without problems (Figure 6.9), and the goal execution ends successfully (Figure 6.10). As in the case of structure sharing, the block associated with the call of $p(a)$ may then be recovered, but the copy stack remains unchanged, because it is updated only upon backtracking.

6.4. Comparisons

Structure sharing and structure copying both lead to the same architecture of the working area in three stacks that are managed in a similar way. Two criteria will then allow us to compare these two methods: the time factor, by measuring the creation and access speed to the instances of terms, and the space factor, by examining the sizes of the *local* and *global stacks* respectively.

6.4.1. Creating and Accessing Terms

Structure sharing favors the creation of new term instances. Indeed, every L3 binding construction is made by simple assignment of the molecule representing the variable (see section 5.2.2). On the other hand, accessing a subterm often requires the traversal of successive references (see section 5.2.2).

Structure copying favors the access to the subterms (see section 5.3.1). On the other hand, creating new term instances requires the copying of the models. This technique also leads to numerous locations being linked among one another and representing the same unbound variable. The access to such a variable is thus found to be slowed slightly.

In terms of speed, structure sharing seems to be more efficient for programs that generate an important number of term instances and that rarely decompose them. Conversely, structure copying will be more appropriate for programs that generate few new terms but work on accessing existent terms. As C. Mellish [Mel82] observed, a "typical" Prolog program usually falls between these two extremes.

Finally, we conclude by observing that a tendency in favor of structure copying has appeared since 1983–84. Indeed, this term representation assures a better "clustering for the addresses," through the way it manages the bindings being constructed [BBC83]. Consequently, structure copying proves to perform more with respect to time on systems having a virtual memory management by leading to fewer page faults than structure sharing.

6.4.2. Memory Space

The composition of the activation blocks for the *local stack* is identical, except that for structure sharing we can optimize the rerun part of the choice blocks by not holding the top field of the *global stack* (see sections 6.2.4, 6.3.3). On the other hand, the environment sizes are different.

Structure sharing uses a memory space of fixed size regardless (constructing or accessing) of the L3 bindings. The duplication of a clause *cl* comprising n variables always produces an environment of size n spread in nl over the *local stack* and in ng over the *global stack*, where nl and ng are the numbers of local and global variables respectively.

For structure copying, the evaluation is more difficult to achieve. Indeed, it is a function of the number of bindings in construction as well as a function of the

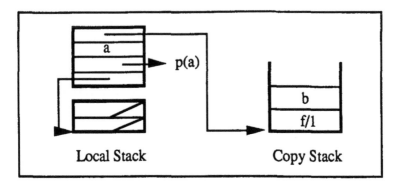

FIG. 6.9. Updating using structure copying (b)

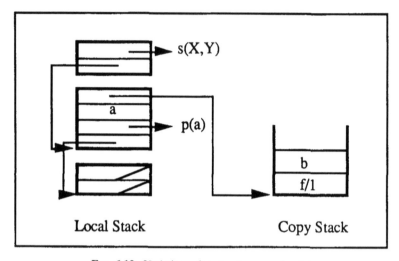

FIG. 6.10. Updating using structure copying (c)

sizes of the copied models of terms. The duplication of a clause *cl* composed of *n* variables produces an environment of size *n* on the *local stack*, plus a space of an a priori undetermined size occupied by the copies on the stack.

Finally, in the context of structure sharing, the distinction between the local and the global variables is made statically. This leads us to consider as global some variables that behave in a local manner (they cannot generate bindings of L3c type). The size of the *global stack* will be found to be uselessly incremented. The *mode declarations* of the Prolog-Dec10 [War77] compiler could be considered as a remedy to this problem.

6.4.3. *Mode Declarations* of Prolog-Dec10

The *mode declarations* of the Prolog-Dec10 compiler allows the restraint of the use of a predicate by specifying the behavior of its arguments. Due to this fact, we can indicate for each argument whether it is an input argument by + or an output argument by –, or if it can act indifferently by ?.

Let us consider the definition that states that an element belongs to a list:

```
member(X,[X|L]).
member(X,[Y|L]):-member(X,L).
```

If we wish to use member/2 only for testing the membership of an already existent list, we should specify that its second argument is an input argument by declaring: |?-mode member(?,+).

If we systematically use the predicate conc/3 to calculate the concatenation of two existent lists, we would declare the first two arguments as input and the third as output: |?- mode conc(+,+,-).

By providing supplementary information about the arguments of each call to a predicate, the *mode declarations* enable the enhancement of the management of the *global stack* and the optimization of the performance of the code generated for the unification.

The *mode declarations* provide a local status to some global variables that do not generate L3c bindings. Indeed, until now we have considered as global every variable liable to generate such a binding (see section 6.2.1). The memory management of the *global stack* is hence improved.

Let us consider the call of a goal b, and let arg be one of its arguments specified as input. During the unification of arg with a structured term te, which figures in the same position at the head of the clause, all the variables of te will be bound to the corresponding subterms of arg (because $args$ is an input argument). Thus the variables of te can no longer produce L3c bindings. These variables can be defined as local if they do not appear in another structured term.

In the case of member/2 without declaration of modes, X and L are global for the fact, while X is local, and Y and L are global for the rule. If we declare the second argument as an input argument, then X and L become local for the fact. Better, because L only appears once, it becomes void, while X is temporary. For the rule, L turns out local and Y void.

In the case of conc/3, defined as:

```
conc([],X,X,).
conc([T|Q],L,[T|R]):-conc(Q,L,R).
|?- mode conc(+,+,-).
```

X remains a temporary variable for the fact. On the other hand Q becomes local with respect to the rule. T remains a global variable, since it does have an occurrence in another structured term not specified as an input argument.

Thereafter, while the use of the *mode declarations* is completely justified for compilation (see chapter 9), it should be simply considered as an enhancement of the memory management in an interpretive approach.

6.4.4. Garbage-Collector on the *Global Stack*

Up to this point we have not considered updating the *global stack* except for backtracking. In the case of deterministic programs that use many structured terms, the stack might suffer an overflow.

Aside from backtracking, the *global stack* may be updated by a garbage-collector [War77], [Bru82b], [ACHS88]. This is achieved in two steps: first, the accessible locations on the *global stack* (or copy stack) are marked by walking through the *local stack*; next, the *global stack* (or copy stack) is compressed by recovering the nonmarked memory locations. But this compression forces us to update all references toward the *global stack* (global-global direction, local-global direction, *trail*-global direction). The implementation of such a garbage-collector forces the development of complex and expensive techniques [Mor78].

In that context, structure copying is more efficient [Bru82b]. Indeed, structure sharing forces the access to a variable through a vector offset. So the entire environment must be kept if a single variable is marked.

Finally, we conclude by observing that the choice of any one of the two representation modes of the terms must not only take into account the inherent requirements of the language development but also consider the host machine characteristics: word length, available memory space, addressing possibilities. We advise the interested reader to refer to [Mel82], where he or she may find a particularly interesting study on this issue.

CHAPTER 7

LAST-CALL OPTIMIZATION

Last-call optimization [Bru80], [Bru82a], [War80] manages specifically the call to the last literal of each rule. If the execution of the father goal has until now been deterministic, then the activation block associated with the call of the last literal replaces the father block. This optimization resembles the iterative management of tail-recursive calls developed in other languages such as Lisp [Gre77], [ASS85] but applies to all last-position calls, recursive or not.

Last-call optimization may be considered as a generalization of deterministic calls updating presented in chapter 6. Indeed, the space allocated for the execution of a deterministic call is recovered no longer upon return but upon the call of its last literal. The *local stack* allocation is then found to be considerably reduced. In particular, tail-recursive calls may, from now on, be executed in a fixed local space.

But this optimization applies only to deterministic situations. In order to be fully efficient, the number of remaining choice points must be reduced to their minimum. Accordingly, clause indexing provides a considerable impact (see chapter 8). Furthermore, this optimization updates only the *local stack*. In order to be fully efficient, its action must be associated with that of a garbage-collector on the *global stack*.

Implementing such an optimization requires a modification in the form of the activation blocks proposed in chapter 6. The disappearance of the father block upon the last literal call causes an irrecoverable loss of information, for the forward process as well as for backtracking.

The last-call optimization was first introduced by M. Bruynooghe [Bru80], for a system using structure copying. D. H. Warren built the first implementation using structure sharing [War79], [War80]. Since the concept and the implementation of this optimization appear to be identical in both cases, our presentation will not rely on any particular representation of terms.

First, we will describe the concept of last-call optimization. Then, we will evaluate its impact by specifying both the role of the clause indexing and the importance of a garbage-collector for the *global stack*.

Subsequently, we will implement this optimization. The appearance of new constraints will lead us to adapt the form of the activation blocks described in chapter 6. We will then reach for a final organization of the blocks suitable for both structure sharing and structure copying. However, two differences will exist for environment management (see section 6.4.2) and for the saving of the arguments of a goal (see section 7.2.3).

The architecture thus obtained is identical to that of the Warren Abstract Machine [War83], which uses structure copying. We will return to this in more detail in chapter 9, when we discuss the compilation stage; then we will propose an interpreter in chapter 12.

7.1. Last-Call Optimization

We will first represent last-call optimization by specifying its application conditions. Next, we make a first evaluation of its effect in terms of memory allocation. Then we show that in order for this optimization to be fully beneficial, it is desirable to have both a clause indexing mechanism [War77] and a garbage-collector for the *global stack* (see section 6.4.4). Finally, a more complete evaluation will be proposed in chapter 8, devoted to the clause indexing mechanism.

7.1.1. Concept

Last-call optimization is a generalization of the updating upon return from deterministic calls, and hence constitutes a space economy measure for the *local stack*.

Its application depends upon the following two conditions: the call must be in the last position, and the execution of the caller (father in terms of the *and/or* tree) must then be deterministic. Under this hypothesis, tail recursions can be henceforth executed in a fixed local space.

7.1.1.1. Presentation

Like other languages such as Lisp [Gre77], [ASS85], and under the hypothesis of determinism, tail-recursive calls are managed in an iterative manner, that is, in a fixed local space.

Consider the following query |?- conc([a,b,c,d,e],[f,g],X). with conc/3 defined as :

```
conc([],X,X).
conc([T|Q],L,[T|R]) :- conc(Q,L,R).
```

The answer to such a query is produced in a fixed local space. Because of determinism, tail-recursive calls are transformed. The conc([b,c,d,e], [f,g],R1) block replaces that of the initial call, and so forth until the conc([],[f,g],R5) call (which creates a choice point) is reached. There is hence constantly one block on the *local stack*. Following the update upon return of deterministic calls (see chapter 6), six blocks would have been present on the *local stack*.

Consider the query |?-last(X,[a,b,c,d,e]). where last/2 is defined as:

```
last(X,[X]).
last(X,[_|R]):-last(X,R).
```

The execution is performed again in a fixed local space (i.e., there is always one block on the stack). Had the updating been performed upon return from deterministic calls, a stacking of six blocks would have been necessary.

This optimization is applicable not only on tail recursive calls but on every call in last position.

```
p1 :- p2.
p2 :- p3.
p3 :- p4.
    ...
pn-2 :- pn-1.
pn-1 :- pn.
pn.
```

The query |?-p1. leads also to a fixed allocation for a single block on the *local stack*. Upon the call of p2, and because of determinism, its block replaces that of p1. The process continues, each pi call replacing the block of pi-1. In the case of updating upon return from deterministic calls, n consecutive blocks would have been necessary.

Last-call optimization can hence be considered as an extension of the iterative management of the tail recursions [Boi85]. However, this optimization is closely tied to determinism, as its application conditions show.

7.1.1.2. Two Application Conditions

Let g be a literal in the body of a rule cl, which has been activated by the call of a goal $gfather$. Two conditions are necessary in order to apply last-call optimization:

1. last position: g is the last literal of cl; and
2. determinism: the execution of the caller $gfather$ did not, until now, generate any choice point.

First of all, the activation block of the caller $gfather$ must be located on the top of the *local stack*, in order to allow its replacement by the activation block consecutive to the call to g. This will be the case if and only if no choice point has been generated. Under this hypothesis, all the left-hand sibling literals of

g have been executed in a deterministic manner, and their blocks have thus vanished from the *local stack*.

Finally, the activation block of the caller *g father* must be deterministic in order to be replaced: the brutal recovery of a choice block would modify the backtracking behavior.

7.1.2. First Evaluation

Last-call optimization constitutes a space economy measure for *local stack* allocation; however, two facts are evident:

1. its application is heavily dependent upon the deterministic behavior of the execution of a Prolog goal, as the conditions to be satisfied show; and

2. it does not generate any *global stack* updating.

Therefore, any measure that reinforces determinism will favor the application of such an optimization, thus enhancing the management of the *local stack*. Moreover, even if programs that need an important local space can now be executed, we should not forget that the global space allocated for representing structured terms is not being recovered.

7.1.2.1. Reinforcing Determinism

Let us reconsider the two examples proposed in section 7.1.1, and let us reverse the clauses ordering inside each of the two sets. Last-call optimization does not apply because each last recursive call now generates a choice point. In both cases, a stacking of six blocks is performed on the *local stack*, even though the deterministic aspect of both is not questioned.

To solve this problem, we can first consider imposing determinism in a brutal manner by adding a *cut* just before the call of the last literal. But this might indeed modify the general behavior of the procedure and, especially, help in the disappearance of the initial declarative aspect. A natural solution, such as clause indexing, is far preferable.

Clause indexing [War77], [War83] reinforces the deterministic behavior of the execution of certain goals by diminishing, upon each call, the figured size of the clauses set. The creation of unnecessary choice points is avoided, thus enabling application of the optimization. Refer to chapter 8 for a presentation of clause indexing and for a discussion of how it relates to last-call optimization.

7.1.2.2. Garbage-Collecting the *Global Stack*

Last-call optimization is solely a space economy measure for the *local stack*. Indeed, in the absence of a garbage-collector, the *global stack* cannot be updated without backtracking (see section 6.4.4).

Consequently, one should be cautious about tail-recursive loops. Let us consider the design of a top-level loop:

```
myloop :- read(X), manage(X,Y), write(Y),
    myloop.
loop :- repeat,
    read(X), manage(X,Y), write(Y),
    fail.
repeat.
repeat :- repeat.
```

In the case of the `myloop/0` predicate, the necessary global space for the execution of `manage(X,Y)` is not recovered between two consecutive calls to `myloop/0`. The predicate `loop/0` achieves a similar treatment, by enforcing backtracking endlessly with `repeat/0`. The *global stack* is here updated upon each loop traversal.

In general, the recursive version must be favored [SS87], but in the absence of a garbage-collector the `repeat-fail` version becomes necessary for writing large systems.

7.2. Implementation

The implementation of last-call optimization requires the modification of the activation blocks form described in chapter 6. Indeed, upon the last literal call, the father block vanishes from the *local stack*. It then becomes necessary to manage differently the information accessed through the initial block during the forward process and during backtracking.

First, we will discuss the forward process by introducing a new notion of continuation and by solving the problem caused by the accessing of the last goal arguments. Afterward, we will examine the consequences of our choices, on the implementation of backtracking. Then, we will propose a new organization of the activation blocks that would be valid for the two modes of term representation.

7.2.1. Forward Process

During the forward process, the father block assures two essential functions:

1. it allows, through its continuation part, access to the remaining sequence of goals still to be proved; and
2. it provides to each of its children the environment in which they must be executed.

Since the father block vanishes upon the last literal call, it becomes necessary to extract these two pieces of information from the *local stack*. Hence, two modifications [War80] must be made to the composition of the activation blocks described in chapter 6:

1. the continuation part will directly designate the next sequence of goals to be proved; and

2. the arguments of a goal will be saved, before unification, in special registers.

7.2.2. Continuation

The continuation part of an activation block will still have two fields (see sections 6.2.4, 6.3.3) but will directly designate the next sequence of goals to be proved [War80], [War83] instead of systematically designating the right-hand sibling literals through the father block:

- field CP: next sequence of goals to prove,
- field CL: local block of CP.

Consequently, the continuation part coding no longer respects the initial *and/or* tree structure (see section 4.2.5). On the other hand, the access to the next sequence of goals to be proved is enhanced. During a last position call, it is unnecessary to traverse different parent blocks to determine the next sequence of goals to be proved.

7.2.3. Saving a Goal Arguments

A goal is no longer able to access its arguments through its father block. Consequently, its arguments must be saved, upon its call, in special Ai registers [War80]. Moreover, this saving must not leave any reference toward the initial block to be recovered [Bru82a], [War80].

After having explained the concept of the saving a goal arguments, we will describe the problem caused by the existence of eventual references toward the father block, which is to disappear from the *local stack*. We will then characterize the variables that generate such accesses by defining the concept of unsafe variables. Afterward, we will present the solutions for structure sharing [War80] and for structure copying [Clo85], [War83].

7.2.3.1. The Problem

Upon a goal call, its arguments are saved in registers before unification. This saving is performed by consulting the caller environment (father). This information being extracted, the latter block may thus disappear from the *local stack* by applying last-call optimization. This saving must not leave any reference to the environment of the block just destroyed.

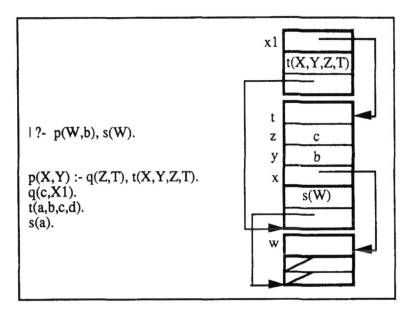

FIG. 7.1. Unsafe variables (a)

Let us consider an example involving no structured terms, in order to be able to work, for now, regardless of a particular term representation.

Figure 7.1 describes the bindings state after the unification between the goal $q(Z, T)$ and the fact $q(c, X1)$. Three local blocks have been allocated. The first one is associated with the query and has an environment of size 1, holding W. The second block is issued from the call $p(W, b)$ (four variables occur in the rule defining $p/2$). Finally, the third block is associated with the call $q(Z, T)$.

Upon the return of $q(Z, T)$, its block disappears from the *local stack*. Last-call optimization is thus applied to the goal $t(X, Y, Z, T)$. Before updating the *local stack*, we should save the arguments of this call in the corresponding registers. These registers are called Ai [War80], [War83], where each Ai maps to the ith argument of the call.

Upon the call to $t(X, Y, Z, T)$, each Ai register is loaded with the corresponding argument value (see Figure 7.2). There is no problem for the first three arguments. Indeed, the dereferenced values of the first three arguments lead to two constants (b and c) and to a variable held by a prior block (W). On the other hand, because T is still in its initial unbound state, the $A4$ register designates a memory location in the father block. Consequently, it is not possible to optimize the last call. Indeed, the replacement of the top block by that of $t(X, Y, Z, T)$ would produce a dangling reference.

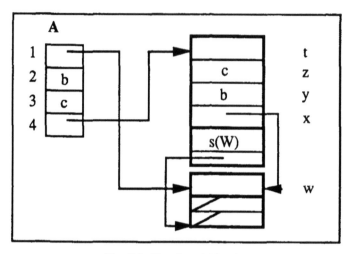

FIG. 7.2. Unsafe variables (b)

7.2.3.2. Unsafe Variables

Such accesses are generated by the free variables that reference the local part of the father block environment, upon the call of the last child literal. Consequently, we say that a variable X, occurring in the last literal of a clause cl, is an unsafe variable if and only if it satisfies one of the following two conditions:

1. X is in its initial unbound state,
2. X is bound to another unbound variable Y of the same environment.

The problem caused by the unsafe variables is solved, either for structure sharing [War80] or for structure copying [War83], by allocating these variables on the *global stack*. Accordingly, the loading of the Ai registers can no longer create references toward the local block that can then be reused.

These two approaches are not strictly equivalent from the point of view of the *global stack* allocation: since the distinction is made in a static manner for structure sharing, every variable that might be unsafe will be made global. On the other hand, for structure copying, the unsafe variables will be dynamically determined and consequently reallocated on the copy stack.

7.2.3.3. Solution for Structure Sharing

The potentially unsafe variables can be statically determined [War80]: these are the local ones that occur in the last literal and do not appear in the head of the rule.

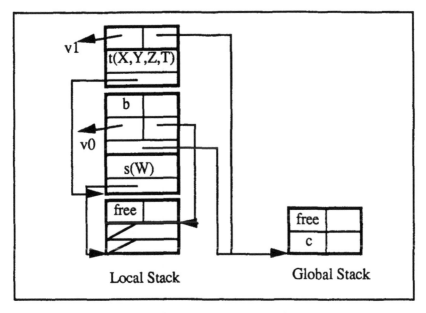

FIG. 7.3. Solution for structure sharing (a)

Indeed, let us first observe that the problem is inherent to local variables, global variables being allocated aside. Furthermore, every local variable occurring in the head of the rule is inevitably bound during unification with the calling goal. Indeed, such a variable occurs at level 1 (see section 6.2.1) and is older than any goal argument. The only variables that are likely to cause a problem are thus the local variables that are found in the last goal and do not occur in the head of the rule. Since in practice the number of these variables is relatively small, D. H. Warren [War80] proposes to make such local variables global.

The initial concept of global variables (see section 6.2.1) is then extended: a variable of a clause cl is now global if and only if it occurs in a structured term or if it occurs in the last literal of cl, without occurring in the head of cl.

Let us consider the previous example (Figure 7.1). Variables Z and T that occur in the rule defining the predicate $p/2$ become global. The bindings state of Figure 7.3 corresponds to the situation previously described in Figure 7.1.

Let us analyze how the arguments of $t(X, Y, Z, T)$ are saved (Figure 7.4). Given the global status of T, no reference to the local block that is to disappear is thereafter performed by the Ai. The optimization is applied by replacing the block on the top of the stack, by that of $t(X, Y, Z, T)$. Afterward, the unification between the Ai registers and the parameters of the fact $t(a, b, c, d)$ succeeds by

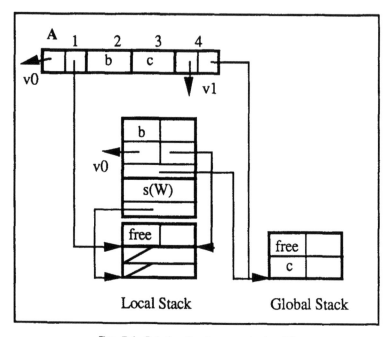

FIG. 7.4. Solution for structure sharing (b)

binding W and T to the constants a and b respectively (see Figure 7.5). The proof ends successfully by applying the optimization again, upon the call of $s(W)$.

This static distinction will result in making global all variables that might be unsafe, not only those that are effectively so. In the previous example, a global status was given to Z even though it is found to be bound to the constant c, upon the saving of the arguments of $t(X, Y, Z, T)$.

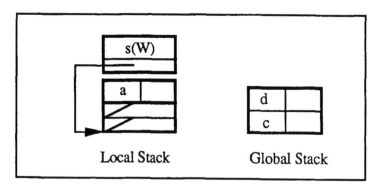

FIG. 7.5. Solution for structure sharing (c)

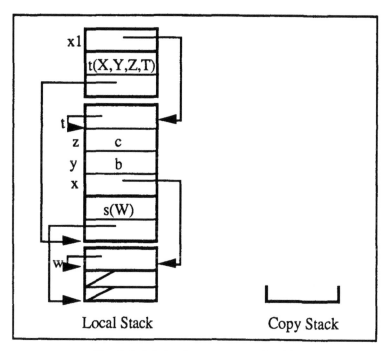

FIG. 7.6. Solution for structure copying (a)

We find here again the characteristics of the static allocation inherent to structure sharing, already discussed in section 6.4.2: every variable that might cause a problem is considered global.

7.2.3.4. Solution for Structure Copying

Unsafe variables are determined dynamically, and consequently reallocated on the copy stack, during the saving of the arguments of the last position goal [War83], [Clo85]. Accordingly, no more references toward the local part of an environment that is to disappear will subsist.

Let X be the dereferenced value of a variable that stands as an argument in a last position goal b. If X is an unbound variable held by the *local stack*, and if X belongs to the father environment of b, then X is an unsafe variable. In this case, saving this argument consists in first generating a new location associated with a free variable at the top of the copy stack; and second, in loading Ai with this value and still making the variable X designate this location. Moreover, the reallocation of X on the heap is made in a manner similar to that of a free variable during the copying of a structured term model (see section 5.3.3).

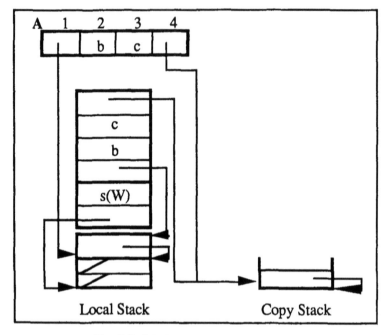

FIG. 7.7. Solution for structure copying (b)

Let us reconsider the example in Figure 7.1, and let us examine (see Figure 7.6) the situation equivalent to that of Figure 7.3. The copy stack is empty, for there is no structured term being used. Upon the call of the goal $t(X, Y, Z, T)$, its arguments are loaded in the Ai registers (see Figure 7.7).

Saving the first three arguments is straightforward. The fourth requires the reallocation of the free variable T. A new location is created on the copy stack, and T references this location. Figure 7.4 describes an identical solution for structure sharing.

The *local stack* is hence updated and the unification then succeeds by binding W and T respectively to the constants a and d (see Figure 7.8). The proof ends successfully by applying the optimization to the goal $s(W)$. Unlike the solution for structure sharing, the variable T, which is the only unsafe one, is allocated on the heap.

7.2.3.5. Loading a Structured Argument

In order to focus specifically on the problem of unsafe variables and on the solutions hence designed, we have disregarded until now the structured terms processing.

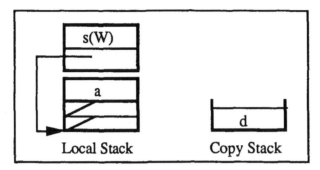

FIG. 7.8. Solution for structure copying (c)

For structure sharing, each Ai register is a molecule. The saving of a structured term te, to be considered in the global environment eg, is achieved by loading the Ai register according to the molecule (te, eg).

For structure copying, each model of a structured term in an argument position is first copied on the top of the heap. The Ai register is then loaded by a reference toward the new term instance thus created.

7.2.4. Backtracking

The modifications brought to the concept of continuation (see section 7.2.2) will bring changes to the organization of the rerun part of the choice blocks. Indeed, these use their continuation part to restart the forward process. Because this access is only correct if continuations respect the initial structure of the *and/or* tree (see section 4.2.4), the choice blocks architecture must consequently be modified.

The choice blocks composition previously proposed (see section 6.2.4 for structure sharing and section 6.3.3 for structure copying) uses the continuation part to restart the forward process. The CP and CL fields allow, upon backtracking, the recovery of the goal instance to be reproved (see section 4.2.4).

For the new concept of continuation, such an access is no longer valid. Let us consider a last position call that generates a choice point. The continuation part of its activation block will directly designate the next sequence of goals to be proved (CP) and the associated local block (CL).

We should thus save an instance of the goal in order to be able to restart the forward process. Accordingly, we save on the *local stack* the n (goal arity) Ai registers already containing its arguments. Consequently, each rerun part of a choice block must have a supplementary A field, consisting of n locations intended to receive the arguments values.

7.2.5. Activation Blocks

We will describe a first organization satisfying the requirements mentioned until now. Afterward, we will bring a modification that enhances the last-call optimization behavior in the case where the father block is a choice block located on the top of the *local stack*.

7.2.5.1. First Organization

A deterministic block always has two parts: continuation and environment. The former is made up of two fields, CP and CL, respectively designating the next sequence of goals to be proved and the associated local block (see section 7.2.2). Environments remain unchanged, both for structure sharing (see section 6.2.3) and for structure copying (see section 6.3.2).

Each choice block has a rerun part that is now composed of five fields:

- the BP, BL, and TR fields respectively designate the remaining clauses set, the previous choice block, and the top of the *trail*;

- the BG field describes the top of the *global stack*. In structure sharing, this information is not kept because it is accessible via the environment part (see section 6.2.4); and

- the A field is made of n consecutive locations holding the n arguments of the goal that has generated the choice point.

Following this form, last-call optimization applies under the two conditions stated in section 7.1.1.

7.2.5.2. An Optimization

Let *lastg* be the last literal of a clause *cl* activated by the call of a goal *father*. If *father* generates a choice block, and if the left-hand siblings of *lastg* are deterministic, no update is performed (see section 7.1.1).

However, the environment part of *father* block has become unnecessary. Indeed, *lastg* arguments being saved, the environment will no longer be accessed. On the other hand, the continuation part of *father* should not disappear: upon each backtracking, it indicates the next sequence of goals to be proved.

We can then consider making each rerun part independent by adding two new fields (BCL, BCP) that save the continuation. In this case, we can recover the continuation and environment parts as for a deterministic block. Indeed, all the necessary information to restart the forward process has been saved in the rerun part, which is composed of seven fields: A, BCP, BCL, BG, BL, BP, and TR.

7.2.5.3. Final Organization

The composition of the activation blocks is thus as follows, regardless of a particular term representation:

- deterministic case:
    1. a continuation part (CP,CL),
    2. an environment part;
- choice case:
    1. a rerun part composed of seven fields (A, BCP, BCL, BG, BL, BP, and TR),
    2. a continuation part (identical to the deterministic case),
    3. an environment part (identical to the deterministic case).

Let us now examine the differences inherent to the two modes of term representation. For structure sharing, each environment is divided into a local and a global part. The problem of unsafe variables is statically solved by providing them a global status. For structure copying, each environment is further divided in two parts, but the global one is dynamically allocated as copyings are performed. Unsafe variables are dynamically reallocated on the heap.

We do find once again the management difference between the two term representation modes (see section 6.4.2): a static approach for structure sharing and a dynamic one for structure copying.

7.2.6. Conclusion

The implementation of last-call optimization that we have just described consists of updating the *local stack* before the unification of the goal in the last position. This process is made possible by saving the arguments of the last goal in the *Ai* registers and by modifying the composition of the activation blocks.

Another solution consists of applying the optimization after the unification of the arguments of the goal in last position. In this case, the father block and the last literal block are, for an instant, both present on the *local stack*. The application of the last-call optimization consists in replacing the father block environment by that of the last child block, and by updating the *local stack* accordingly. For a description and an implementation of this method, we advise the reader to refer to [KS85].

We have chosen the first approach because it is the most commonly used. Moreover, it fits our initial objective: to evolve naturally toward the compilation stage (see chapter 9), for which Warren's Abstract Machine [War83] has been chosen as the reference. Indeed, in the context of structure copying, it uses the activation blocks architecture just described (see section 7.2.5). However, the implemented technique for updating the *local stack*—known as trimming—generalizes the last-call optimization presented above. We will describe the trimming principle and its implementation in section 9.1.3.

CHAPTER 8

CLAUSE INDEXING

The purpose of clause indexing [War77], [War83], [BBC83], [BBCT86] is to diminish, upon every goal call, the set size of selectable clauses, thus avoiding useless unification attempts and unwanted choice points creation.

Indeed, upon a goal call, it is common practice (see section 2.2.2) to consider selectable all the clauses occurring in the set. However, quite often only one part among all selectable clauses will satisfy unification with the calling goal. We can thus try initially to disregard some of these.

Accordingly, every set of clauses is organized (divided) into subsets. Upon every call of a goal its arguments' type (constant, variable, list, or functional term) is analyzed. The corresponding subset is then accessed. In order for this analysis to be efficient, and for the subset not to lead to an intricate representation, indexing is performed only on the first argument [War77], [War83], [BBC83], [BBCT86].

Clause indexing has a twofold benefit. First, it avoids useless unification attempts, thus enhancing the timing performances. Second, it optimizes memory management by enforcing determinism. The economical measures for *local stack* allocation, which we have presented in chapters 6 and 7, will now be applied more frequently.

After studying its concept, we will evaluate the benefit of clause indexing whether the *local stack* is updated upon returns from deterministic calls (see chapter 6) or by the last-call optimization (see chapter 7).

8.1. The Concept

The purpose of clause indexing is to avoid unification attempts whose failure is a priori known. This is done by analyzing the goal arguments type and by retaining the clauses that might satisfy the unification. In order to make this selection simple and efficient, we usually [War77], [War83] consider the goal first argument only.

Implementing the indexing mechanism leads to the modification of the initial structure of each set of clauses. Indeed, if a linear representation (e.g., a simple list of clauses) were sufficient until now, we should henceforth divide every set into subsets according to the type of the first parameter of the clauses' heads. Finally, this new organization must respect the clauses' initial sequence in order to keep the search rule unchanged (see section 2.2.2).

We will describe a first-level indexing based on a simple analysis of the type of goal first argument and then fine-tune this approach by proposing a second-level indexing.

8.1.1. First-Level Indexing

For a better understanding of the organization of the sets of clauses, let us first study how selection is performed. We usually distinguish [War83], [BBCT86] four primary keys depending on the type of term addressed: constant, variable, list, or functional.

Based upon the goal first argument analysis, clause selection is made in the following manner: let $A1$ be the dereferenced value of the goal first argument:

- if $A1$ is a variable, the whole set is to be retained;

- if $A1$ is a constant, then only clauses having a variable or a constant as first parameter of their head should be considered;

- if $A1$ is a list (a functional term), only clauses having a list (a functional term), or a variable, as the first parameter of their head are to be selected.

Thus, the clauses' set organization must allow such an access, while still respecting the clauses' initial sequence. We will propose an implementation in section 11.1.3 that will be suitable for our three implementations: Mini-Cprolog (see chapter 11), Mini-WAM (see chapter 12), and Mini-Prolog-II (see chapter 13). We will describe the specific aspects of the compilation of a set of clauses in section 9.2

As an example, let us borrow from D. H. Warren [War77] the definition of the predicate call/1:

```
call(or(X,Y)) :- call(X).             % c1
call(or(X,Y)) :- call(Y).             % c2
call(and(X,Y)) :- call(X), call(Y).   % c3
call(trace) :- trace.                 % c4
call(nl) :- nl.                       % c5
call(X) :- builtin(X).                % c6
call(repeat).                         % c7
call(repeat) :- call(repeat).         % c8
```

Four subsets are created in provision for further accesses. In the first subset, associated with the case of a variable, all clauses are satisfactory. In the case of a constant, the subset holds clauses $c4$, $c5$, $c6$, $c7$, and $c8$. For a list, only clause $c6$ is to be retained. Finally, for a functional term, only clauses $c1$, $c2$, $c3$, and $c6$ are considered. A clause that has a variable as its first parameter in its head belongs to all the subsets. This is the case for clause $c6$.

8.1.2. Second-Level Indexing

This first indexing scheme can be enhanced by a second-level indexing [War83], [BBCT86], except for the case of a variable where the whole initial set is to be considered:

- constants are indexed by their names,
- functional terms are discriminated according to their functor, and
- lists should be distinguished as empty and nonempty.

For the predicate `call/1`, the second-level indexing leads to the following situation:

- Variable: $c1$, $c2$, $c3$, $c4$, $c5$, $c6$, $c7$, $c8$.
- Constant:
  - `trace`: $c4$, $c6$.
  - `nl`: $c5$, $c6$.
  - `repeat`: $c6$, $c7$, $c8$.
- Functional term:
  - `or/2`: $c1$, $c2$,$c6$.
  - `and/2`: $c3$, $c6$.
- List:
  - empty list: $c6$.
  - nonempty list: $c6$.

Finally, we observe that the distinction between empty and nonempty lists is heavily dependent upon the representation chosen for the empty list. If the latter is represented (as is often the case) by a particular constant, for example the atom `nil`, then the empty list case is identical to that of a constant, the list indicator being reserved for nonempty lists.

8.2. *Local Stack* Enhancement

By shrinking the size of the retained sets, clause indexing reduces the number of choice points allocated in the *local stack*. This reenforcement of determinism enhances the *local stack* management, whether updating is performed only upon return from deterministic calls or achieved through last-call optimization.

After giving a first example, we will evaluate more precisely the benefit of clause indexing. We will therefore compare the *local stack* allocation in light of the two updating techniques, first without using indexing and next while using

it. To achieve this evaluation we are going to consider the classical example of
the naive reverse of a list.

8.2.1. First example

The problem mentioned in section 7.1.2, concerning the clause ordering in the
definition of the predicate conc/3 and last/2, is no longer relevant. Upon
every call, the retained set is reduced to a single clause: the rule for the first
five calls and the fact for the last one. Note that in this case the allocation of
the final choice point is also avoided (see section 7.1.1).

This is crucial. The allocation of an unnecessary choice point forbids any
further attempt to update the *local stack*, whether upon return from deterministic
calls or through last-call optimization. The behavior of the naive reverse of a
list illustrates this observation perfectly.

8.2.2. Naive Reverse

We will study the *local stack* allocation, after evaluating the number of necessary
calls in function of the size of the list to be reversed. This will be achieved by
comparing, first without indexing, last-call optimization versus updating upon
return from deterministic calls. We will show afterward the benefit of clause
indexing.

8.2.2.1. Number of Calls

Let us consider the definition of the naive reverse of a list:

```
nrev([],[]).
nrev([T|Q],R) :- nrev(Q,R1),
    conc(R1,[T],R).
conc([],X,X).
conc([T|Q],L,[T|R]) :- conc(Q,L,R).
```

The number of calls for reversing a list of size n satisfies the following
recursive relation:

$$u_0 = 1; u_n = 1 + u_{n-1} + n; i.e, u_n = (n+1)(n+2)/2. \qquad (8.1)$$

Upon each call of nrev/2 on a nonempty list, the tail Q of this list must be
recursively reversed and the initial list head T must then be inserted at the end
of the resulting reversed list R1. The number of calls to conc/3 is equal to the
size $n-1$ of its first argument plus 1 (for the empty list). Thus, the total number
of calls is quadratic with respect to the size of the list to be reversed.

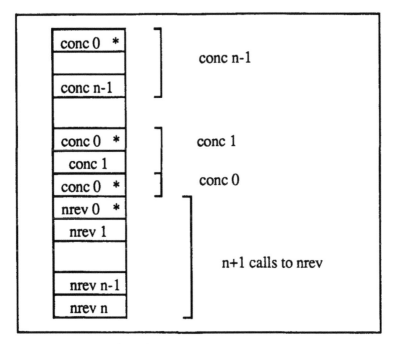

FIG. 8.1. Updating upon return from deterministic calls

8.2.2.2. Evaluation without Indexing

Let us focus first on the order of the calls. For a list of size n, the $n + 1$ calls to nrev/2 are first executed. Each call to nrev/2 with a list of size i then starts the execution of conc/3 with a list of size $i - 1$ as its first argument. Without indexing, each call relative to the empty list (nrev/2 or conc/3) allocates a choice point since in both cases a rule remains to be considered.

In updating only upon return from deterministic calls, all the allocated blocks are found on the *local stack* (see Figure 8.1) where choice points are marked by $*$). The blocks allocation is thus quadratic with respect to the list size. Upon return from $nrev_n$, all blocks remain on the *local stack*, because of the existing choice points.

Last-call optimization (see Figure 8.2) will allow deterministic calls to conc/3 to be executed in a constant local space. Indeed, every recursive call to conc/3 will be optimized, and the last one (empty list) will generate a choice point.

Blocks allocation becomes linear with respect to the list size. Again, upon return from $nrev_n$, all blocks remain present, here too, on the *local stack*.

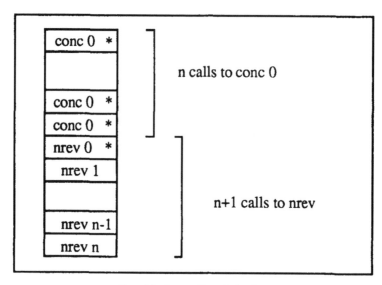

FIG. 8.2. Last-call optimization

8.2.2.3. Evaluation with Indexing

Indexing will turn any call that involves the empty list deterministic and will therefore favor the *local stack* updating, either upon return from deterministic call or by last-call optimization.

Let us study the *local stack* behavior when updating upon return from deterministic calls. Figure 8.3 describes the situation just before the return of $nrev_i$.

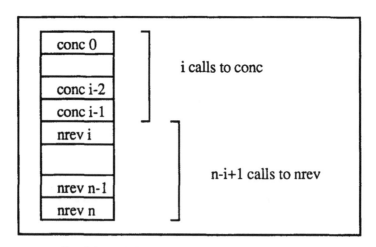

FIG. 8.3. Updating upon return from deterministic calls

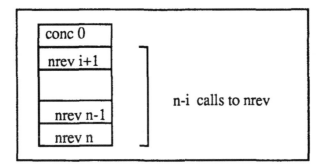

FIG. 8.4. Last-call optimization

This call to nrev/2 has triggered the execution of the recursive call $nrev_{i-1}$. Since the execution of the latter is deterministic, the allocated space is recovered as soon as it returns. Next, $conc_{i-1}$ is executed, creating i deterministic blocks. The *local stack* size thus ranges between $n - i + 1$ and $n + 1$. Upon $nrev_n$ return, the *local stack* recovers its initial status prior to the call.

Because of determinism, last-call optimization will apply systematically. Upon return from $nrev_i$, the state of the *local stack* is as described in Figure 8.4.

Last-call optimization is applied on $conc_{i-1}$ call. Its block replaces that of $nrev_i$. Last-call optimization is then applied as before to deterministic calls to conc/3 triggered by $conc_{i-1}$. The *local stack* size grows to $n + 1$ (the $n + 1$ initial calls to $nrev$), then decreases in a $n + 1 - i$ fashion. The *local stack* recovers its initial state upon $nrev_n$ return.

8.2.2.4. Comparison

Let us summarize these results in two tables: one for the updating upon return from deterministic calls (see Figure 8.5) and the other for the last-call optimization (see Figure 8.6).

Because clause indexing avoids the allocation of all useless choice points, the local space needed for the execution is found to be considerably smaller. Most importantly, since the proof has become totally deterministic, the *local stack* recovers its initial status prior to the call upon $nrev_n$ return.

Of course, this version of list reversing is far from being as efficient in terms of time and space saving as the tail-recursive one using an accumulator [War77]:

```
nrev(X,Y) :- nrev1(X,Y,[]).
nrev1([],R,R).
nrev1([T|Q],R,R0) :- nrev1(Q,R,[T|R0]).
```

The reversing comes of linear complexity and can be performed in a constant local space (one block only) by applying last-call optimization.

without Indexing		with Indexing	
max. size	after return	max. size	after return
$\dfrac{(n+1)(n+2)}{2}$	$\dfrac{(n+1)(n+2)}{2}$	n+1 between n-i+1 et n+1 for $0 \le i \le n$	0

FIG. 8.5. Updating upon return from deterministic calls

without Indexing		with Indexing	
max. size	after return	max. size	after return
2n +1	2n +1	n+1 incr. to n+1 decr. like n+1-i for $0 \le i \le n$	0

FIG. 8.6. Last-call optimization

144

CHAPTER 9

COMPILATION OF PROLOG

The first Prolog compiler (Prolog-Dec10) was developed at the University of Edinburgh by D. H. Warren in 1977. Since then, much research has been conducted all over the world in this area [War80], [War83], [Bal83], [BBC83], [BBCT86], [Clo85]. Currently the overwhelming majority of commercialized Prolog systems embed a compiler.

Most of the proposed solutions are based on the concept of a virtual machine in order to achieve portability. However, as D. H. Warren points out [War83], the intermediate code thus generated can be processed in different ways: by using an emulator [Qui86], by generating machine code [Car86], by microprogramming [GMP87], or directly by a dedicated processor [Tic83], [DDP84], [NN87]. [MT87] proposes a comparative study of these last two implementation modes.

We will retain the approach of designing a compiler based on the concept of a virtual machine. For that, we will essentially study the proposal made in 1983 by D. H. Warren, which is known under the name WAM (Warren's Abstract Machine). Indeed, the WAM has been abundantly referred to in other works and is the de facto standard for implementing Prolog compilers.

Our presentation will consist of two parts. We will first describe the implementation choices made by the WAM and will justify them through the review of the problems and solutions we have exposed in the five previous chapters. We will then deal with the aspects specific to compilation. For that, we will distinguish two compilation levels: the set of clauses and the clause itself. We will conclude by focusing on the extensions and enhancements that have been made to the initial definition of the WAM.

Compiling a set of clauses consists in translating it into a sequence of instructions that achieves the indexing of its clauses, together with its backtracking management.

Compiling a clause consists in generating the instructions that implement its duplication, the unification of its head, and the call to the various goals held in its body.

We will describe throughout our presentation the five main categories of primitives of the WAM's instruction set:

1. Clause indexing and backtracking management: *indexing*;

2. Forward process: *control*;

3. Unification with a clause's head: *get, unify*;

4. Registers loading upon the call of a goal: *put, unify*.

9.1. Characteristics of the WAM

The WAM uses structure copying, and its working area is classically divided in three stacks: the *local stack*, the heap, and the *trail*. The *local stack* is updated using trimming, which constitutes a generalization of last-call optimization. Finally, the WAM implements the two levels of clause indexing presented in section 8.1.

9.1.1. Terms Representation by Structure Copying

The WAM uses structure copying. The terms representation is achieved by using tagged words that describe the object's type (see Figure 9.1). Unbound variables are self-referenced.

For the structured terms, lists and functional terms are differentiated (see Figure 9.2). Besides a more compact coding (it is useless to support the functional symbol in the case of a list), this differentiation will allow a more specific management of the lists during unification, thus enhancing timing performances. Finally, the empty list is represented by the *nil* constant.

9.1.2. Architecture

The working area of the WAM is organized into three stacks (*trail, local stack,* and heap). The *local stack* is updated using trimming that consists in a progressive application of the last-call optimization. After describing its memory organizational, we will describe the structure of the WAM's activation blocks.

9.1.2.1. Architecture of the Working Area

Besides the code zone where the Prolog procedures are found, the WAM possesses the three usual working stacks organized as depicted in Figure 9.3.

Let us recall that the only organization constraint is that the heap be located below the *local stack* in order to simplify the binding algorithm's task of respecting creation times (see section 6.3.4). The Push Down List (PDL) is a memory area used for unification.

Let us explain the purpose of the WAM's registers with respect to those we have chosen for our description (see chapter 7) and for our implementations, especially those of chapter 12:

- Continuation registers:
  - *CP*: *CP* (next sequence of goals to prove),

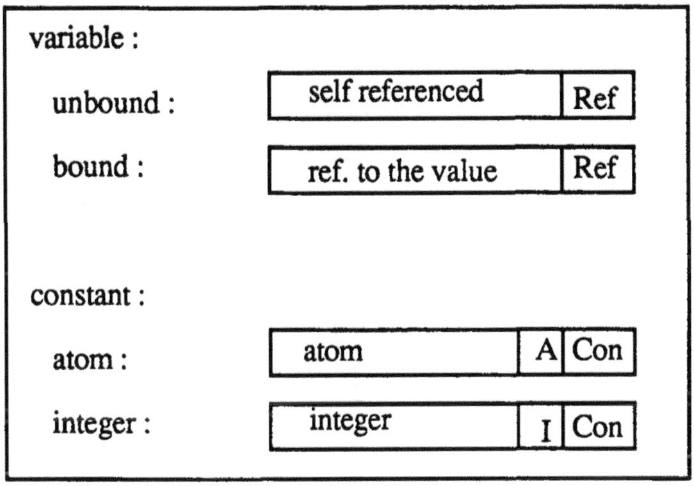

```
variable :

    unbound :        | self referenced    | Ref |

    bound :          | ref. to the value  | Ref |


constant :

    atom :           | atom          | A | Con |

    integer :        | integer       | I | Con |
```

FIG. **9.1.** Simple terms representation

- *CL*: *E* (local *CP* block);
- Backtracking registers:
  - *BL*: *B* (last choice block),
  - *BG*: *HB* (top of the heap associated with *B*);
- Top of stacks:
  - *L*: *A* (top of the *local stack*),
  - *G*: *H* (top of the heap),
  - *TR*: *TR* (top of the *trail*);
- Management of the current goal's call:
  - *Ai*: *Ai* (saving of the current goal's arguments),
  - *PC*: *P* (next goal to be proved).

Apart from their names, these registers carry the same functions as those assigned to them in section 2.1 of chapter 12.

9.1.2.2. Structure of the Activation Blocks

The structure of the WAM's activation blocks is the same as the one retained in section 7.2.5, the rerun parts of the choice blocks being totally independent. This approach will allow the separate management of both the forward and backtracking processes. We will discuss this in section 9.2, which describes the compilation of a set of clauses.

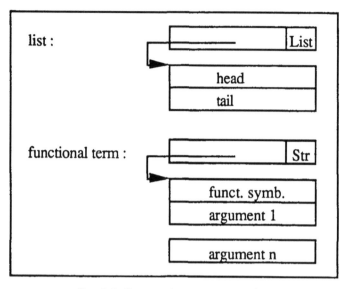

FIG. 9.2. Structured terms representation

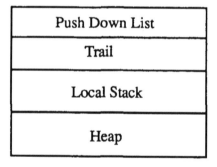

FIG. 9.3. Architecture of the working area

A deterministic block is thus divided into its continuation (CP, CE) and environment parts. A choice block embeds a rerun part for restarting the forward process upon backtracking. It is made up of seven fields: A, TR, BP, B, H, BCP, and BCE. We thus meet, except for their names, the organization described in section 7.2.5.

9.1.3. Trimming

Whereas the *trail* and the heap are conventionally updated upon backtracking, the *local stack* is managed through a generalization of the last-call optimization, called trimming.

9.1.3.1. The Concept of Trimming

Trimming is based on the following observation: taking into account the previous saving mechanism of a goal's arguments (see section 7.2.3), the local memory location associated with a variable is no longer needed after the call of the goal in which its last occurrence figures.

Indeed, if an unbound variable occurs within a structured term itself in an argument position, then the variable will have already been copied (on the heap) during the loading of this argument (see section 7.2.3). Last, we shall extend the concept of unsafe variable to any unbound variable that stands as an argument in the goal where it occurs last.

Consequently, we can then think of recovering the local environment of the father block in a continuous fashion during the proof of its sons and not only at the end, as with last-call optimization.

In practice, the variables of a clause *cl* are numbered first in order to facilitate this updating: the variables with the shortest life span are allocated in the upper part of the environment. Subsequently, for each call of a goal ocurring in the body of *cl*, we associate the size of the environment necessary for the proof's continuation. If the father block is on the top of the stack (no intermediate choice points being allocated), the *local stack* is then updated consequently.

9.1.3.2. Example

Let us consider the following definition of qsort/3:

```
qsort([],R,R).
qsort([X|L],R0,R) :-
    partition(L,X,L1,L2),
    qsort(L1,R0,[X|R1]),
    qsort(L2,R1,R).
```

In this approach the variables of the rule are classified according to their life span, the variable with the highest numbering being first to disappear from the environment. In this case, they have the following order: V1=L2, V2=R1, then R, L1, R0, X, and L. To each goal occurring in the body, we associate the environment's size to be retained:

```
qsort([V6|V7],V5,V3) :-
    partition(V7,V6,V4,V1),    [6]
    qsort(V4,V5,[V6|V2]),      [3]
    qsort(V1,V2,V3).           [0]
```

As soon as partition(L,X,L1,L2) is called (once its arguments are loaded in the *Ai* registers) the *local stack*'s top can be decremented by one, since the variable (L=V7) does not appear in partition(L,X,L1,L2)'s right-hand siblings. Similarly, upon the first recursive call to qsort/3, the space allocated for the V4, V5, and V6 variables can also be recovered, since only V1, V2, and V3 occur in the last goal. Last-call optimization is conservatively applied on what is left of the father block during qsort/3's second recursive call.

9.1.3.3. Trimming and Last-Call Optimization

Trimming can thus be considered as a continuous application of the last-call optimization. Indeed, the size of a father block decreases progressively during its children's calls. Last, the remaining part is recovered upon the call of the last literal, as in the classical last-call optimization.

For trimming to be applied in a continuous fashion or in its final form, the father block should be in the top of the *local stack*, that is, as for the last-call optimization, no intermediate choice points creation should have occurred.

Finally, we should observe that using trimming increases the ratio of unsafe variables (see section 7.2.3). Indeed, the problem of a variable referencing an environment part to be recovered occurs not for the last literal only but upon each call of a goal that tries to reduce the environment size.

We can then ask ourselves about the need to use trimming in the case of a term representation by structure sharing. Indeed, in this case the problem is statically solved (see section 7.2.3) by providing all potentially unsafe variables with a global status. Implementing trimming would thus lead to a reinforcement of the environment's global part at the expense of its local part.

9.1.4. Clauses Indexing

The WAM achieves the two indexing levels described in chapter 8. First, we discriminate according to the type of the first argument: variable, constant, list, or functional term. Afterward, in the case of a constant or a functional term, a second indexing level is performed as described in section 8.1.2.

9.2. Compilation of a Clauses Set

Each set is translated into a sequence of instructions that perform clauses indexing and creation, updating, and removal of the choice points on the *local stack*. Clause indexing and backtracking management are thus highly mixed. We deliberately separated them in our presentation (see chapter 7 for backtracking management and chapter 8 for clauses indexing) and in our implementation (see chapter 12).

We will first describe the first indexing level, focusing on the case where the dereferenced value of $A1$ is a variable. We will then present the second indexing level that deals with constants and functional terms. Afterward, we will give a first example. Finally, we will deal with the clauses having a variable as their head's first parameter.

9.2.1. First-Level Indexing

The subset (variable, constant, list, or functional term) associated with the type of the dereferenced value of $A1$ (see section 8.1.1) is selected. This dispatching is realized by the instruction

```
switch_on_term EtVar, EtConst, EtList, Etfunct
```

which dereferences the $A1$ register and loads the P register with the corresponding label.

In order to take care of the case where $A1$ is a variable, the clauses are linked together as to be sequentially considered according to their order of appearance. To achieve this, the compiled form of each clause is headed by a specialized instruction depending on its position in the set:

- `try_me_else EtiqCL` (precedes the first clause)

  1. creates a rerun part that has `EtiqCl` in its BP field;
  2. updates the B and HB registers;

- `retry_me_else EtiqCl` (precedes each intermediate clause)

  1. restores the environments;
  2. updates the BP field of the last choice block with the `EtiqCL` value;

- `trust-me-else fail` (precedes the last clause)

  1. restores the environments;
  2. pops the last choice point;
  3. updates B and HB;

These three instructions perform creation, modification, and deletion of the choice points. They respectively have the same role as the three functions `push_choix`, `push_bpr` and `pop_choix` described in section 12.2.2.

If the dereferenced value of $A1$ is not a variable the sequence of `try_me_else`, `retry_me_else` and `trust_me_else` instructions is not performed and we directly access the code of the corresponding subset.

9.2.2. Second-Level Indexing

A second-level indexing is used for the constants and the functional terms (see section 8.1.2), the dispatching being performed by the

```
switch_on_constant N, Table
```

and

```
switch_on_structure N, Table
```

instructions, where (N,Table) is a size N structure that enables the mapping, through hashing, of the label of the corresponding subset to a key (identifier or functional symbol).

Choice points management within a subset mapped to one key is autonomous and relies on three specialized instructions. These perform the same process as for the first-level indexing primitives (see section 9.2.1), but this time, the L label designates the current clause.

- `try` L (related to the first clause).
    - creates a rerun part, which holds the next instruction in its BP field;
    - updates the B and HB registers;
    - loads L in the P register;
- `retry` L (related to a clause in the middle of a set).
    - restores the environments;
    - updates the BP field with the value of the next instruction;
    - loads L in the P register;
- `trust` L (corresponds to a clause at the end of the subset)
    - restores the environments;
    - pops the last choice point;
    - updates B and HB;
    - loads L in the P register;

Lists are processed in a similar fashion through the three `try`, `retry`, and `trust` instructions, but with no second-level indexing.

9.2.3. Example

Let us review the example of section 8.1 while still disregarding for the time being the sixth clause, whose first head parameter is a variable:

```
call(or(X,Y)) :- call(X).
call(or(X,Y)) :- call(Y).
call(and(X,Y)) :- call(X),
    call(Y).
call(trace) :- trace.
call(nl) :- nl.
% call(X) :- builtin(X).
call(repeat).
call(repeat) :- call(repeat).
```

The compiled code is then as follows:

```
call/1 :      switch_on_term C1a, L1, fail, L2

L1 :          switch_on_constant 4,(trace:C4; nl:C5; repeat:L4)
L4 :          try C7
              trust C8

L2 :          switch_on_structure 2, (or/2 : L3; and/2 : C3)
L3 :          try C1
              trust C2

C1a :         try_me_else C2a
C1
C2a :         retry_me_else C3a
C2
C3a :         retry_me_else C4a
C3
C4a :         retry_me_else C5a
C4
C5a :         retry_me_else C7a
C5

C7a :         retry_me_else C8a
C7
C8a :         trust_me_else fail
C8
```

If the dereferenced value of $A1$ is an unbound variable, all the clauses are sequentially considered due to the try_me_else, retry_me_else, and trust_me_else instructions. In the case of a constant, selection is made according to switch_on_constant.

For trace/1 and nl/0, the call will be deterministic. However, repeat/0 (defined by two clauses) will require the successive investigation of C7 and C8. For a functional term, we discriminate according to the functors or/2 and and/2, and in the first case, we successively try C1, then C2.

9.2.4. Other Clauses

We still have to investigate the case of clauses having a variable as their first parameter, and which should appear in all subsets. With respect to this matter, the strategy adopted by the WAM consists in splitting the initial set in subsets, depending on whether or not they appear before such clauses. An initial choice point is created, and subsequently each subset is processed independently.

Let us review the previous example by introducing this time the sixth clause:

```
call/1 :      try_me_else C6a
              switch_on_term C1a, L1, fail, L2
L1 :          switch_on_constant 2, (trace : C4; nl : C5)
```

```
L2 :            switch_on_structure 2, (or/2 : L3; and/2 : C3)
L3 :            try C1
                trust C2
C1a :           try_me_else C2a
C1
C2a :           retry_me_else C3a
C2
C3a :           retry_me_else C4a
C3
C4a :           retry_me_else C5a
C4
C5a :           trust_me_else fail
C5
C6a :           retry_me_else L4
C6


L4 :            trust_me else fail
                switch_on_term (C7a, L5, fail, fail)
L5 :            switch_on_constant 1, (repeat : L6)
L6 :            try C7
                trust C8
C7a :           try_me_else C8a
C7
C8a :           trust_me_else fail
C8
```

We first trigger a choice point associated with the sixth clause; then we process the five first clauses separately from the last two. The initial choice point is popped after C6, and a new dispatching is performed on the remaining clauses. For that, we first discriminate upon the dereferenced value of A1. Then, in the case of a constant, we will investigate (if it is repeat/0) successively C7 and C8.

This clauses management—dealing with the clauses that have a variable as their head's first argument—can trigger the allocation of multiple choice points associated with a unique goal. This will be the case, in the previous example, if the dereferenced value of $A1$ leads to a functional term of the form or(t1,t2). Indeed, the first try_me_else will generate a choice point associated with C6. The first try will then create a new one for C2.

9.3. Compilation of a Clause

Compiling a clause consists in generating the instructions that perform its duplication, the unification of its head, and the proof of the different goals held in its body. The choices made by the WAM are inherent to an intensive usage of the Ai registers, in the context of a terms representation by structure copying.

Indeed, the role assigned to the Ai registers will allow the clustering of a particular category of variables: the temporary ones. Three types of clauses, each one having its own compiled form, will emerge from this distinction: facts, rules with a single literal in their body, and the rules with more than one literal in their body. Finally, the unification and the loading of a goal's arguments will also benefit from this approach.

We will first describe the fundamental role assigned to the Ai registers. We will then focus on the control aspect of these three categories of clauses. Unification compilation will be covered next, followed by the goal's arguments loading. Finally, we will conclude by examining the management of temporary variables.

9.3.1. The Role of the Ai Registers

We will start with the interpretive approach. The usage made of the Ai registers in a compilation context will make us distinguish two sorts of variables: the temporary variables and the permanent ones. We will then examine the consequences that result at the control level. Finally, we will conclude by defining the notion of a self-init variable.

9.3.1.1. A First Approach

In an interpretive approach, the management of a clause cl differentiates between two cases depending on whether cl is a fact or a rule, with a special management of the last literal for the latter (last-call optimization). If trimming is applied, the *local stack* updating is repeatedly applied upon each call and in its final form upon the call of the last literal.

Let us consider the call of a goal B (whose arguments are loaded in the Ai registers). Let cl be a candidate rule (having a head h and a body $l1, ..., ln$) for unification. A first compiled form of cl would be as follows:

```
unify h parameters with the Ai registers;
load l1 arguments in the Ai
call l1 applying trimming
.../...
call ln-1 applying trimming
load ln arguments in the Ai
call ln by applying the final step of trimming
```

Of course, trimming (continuous or final) is only applied if the father block is located in the top of the stack, that is, if no intermediate choice points creation occurred.

9.3.1.2. Temporary and Permanent Variables

An X variable belonging to a clause cl will be permanent iff X occurs in at least two goals of cl's body, where the clause's head is considered as "part of the first goal." The other variables in the clause cl will be named temporary.

Thus, considering the rule

```
p(X,Y,Z) :- q(Y,T1,U),
            r(T2,X,U,T2),
            s(X,Z).
```

The U, X, and Z variables are permanent, whereas T1, T2, and Y are temporary. Z is permanent because it occurs in the head and in the third goal, whereas Y is temporary since it only appears in the head and in the first goal.

A permanent variable requires the allocation of a memory location on the *local stack*, whereas a temporary variable can be directly managed by the Ai registers.

Indeed, for a temporary variable X in a clause $cl = h : -h1, \ldots, hn$, three cases might occur:

1. X occurs only in h: X will be directly managed through the Ai's by the unification;

2. X occurs only in one goal hi in cl's body: X will be directly managed through the Ai's during the loading of hi's arguments;

3. X occurs in both h and $h1$: we can merge the unification of h with the loading of $h1$'s arguments.

Consequently, the $(k + n)$ variables that appear in a clause are divided into two categories depending on whether it is necessary to keep them in the environment: the temporary variables are labeled $X1, X2, .., Xk$ and the permanent ones are labeled $Y1, Y2, ..,Yn$.

9.3.1.3. Control Level

According to this classification of variables, only an environment for permanent variables will be allocated during a clause duplication. The temporary variables will be directly managed through the Ai registers, using an adequate numbering (see section 9.3.5).

Thus, each fact or rule with a single literal in its body will contain only temporary variables. Accordingly, they do not require any environment allocation.

9.3.1.4. The Concept of a Self-Init Variable

We can classify the variables occurring in a clause cl in four categories depending on whether their first occurrence appears:

- H1: as a parameter (level 1) in cl's head,
- H2: in a structured term in cl's head,
- B1: as an argument (level 1) of a goal in cl's body,
- B2: in a structured term which is an argument of a goal in cl's body.

We will say that an X variable of a cl clause is self-init iff it satisfies $H1$, $H2$, or $B2$. A self-init variable X is bound, as soon as it first occurs, either by unification ($H1$, $H2$) or by the loading of the structured term in which it occurs ($B2$).

Indeed, with respect to $H1$, X being an initially unbound variable, it will be bound to the term referenced by the corresponding Ai register. For $H2$, let te be the structured term in which X occurs. If te is accessed by constructing (see section 5.1.5), X will be reallocated on the heap, or else X will then be bound to the corresponding subterm already located on the heap (accessing).

For $B2$, let g be the goal where X first occurs: X (in its initial unbound state) will be copied during the loading of g's arguments and will thus be bound to a new free variable on the heap (see section 7.2.3).

Finally, let us point out that if an X variable satisfies $H2$ or $B2$, we are then sure that its first occurrence will trigger a binding to a term held by the heap.

9.3.2. Control

This new variable classification leads us to consider three kind of clauses, depending on the size of their body:

1. the facts,
2. the rules having a single literal in their body, and
3. the rules with at least two literals in their body.

Indeed, because of the absence of permanent variables in the first two categories, no block will be allocated in the top of the *local stack*. For the third category, we will build a block, but we will apply trimming on the environment hence created.

9.3.2.1. Three Kinds of Clauses

A particular compiled form is associated with each of the three clauses categories. The allocation of the environment and rerun parts, together with the control management inherent to the forward process, is performed through the use of a set of specialized instructions (see next section):

1. facts: F

- unify F's parameters against the Ai's;
- `proceed`;

2. rules: `H :- Q.`
   - unify H's arguments against the Ai's;
   - `load Q's arguments`;
   - `execute Q`;

3. rules: `H:- Q1, ..., Qn.  (n>1)`
   - `allocate`;
   - unify H's arguments against the Ai's;
   - `load Q1's arguments`;
   - `call Q1, n1`;
   - `load Q2's arguments`;
   - `../..`
   - `load Qn's arguments`;
   - `deallocate`;
   - `execute Qn`;

No block will be allocated for the first two categories. For the last one, the current continuation is saved and an environment for the permanent variables is allocated. Then, with the body's calls being performed, there is always a continuous trimming first (each `call` instruction with the size of the environment part to save), followed by a final `deallocate`. Of course, trimming is only performed if the father block is in the top of the stack.

9.3.2.2. Control Instructions

- `allocate`: (a rule with more than one literal in its body)
  - allocates a new block on the top of the *local stack* while still performing trimming, if possible;
  - saves the current continuation;
- `deallocate`: (before `execute`)
  - restores the continuation registers;
- `execute Pr`: (Pr is the last literal of a rule)
  - loads Pr in P;
- `proceed`: (after a fact)
  - loads CP in P;

- call Proc,N: (Proc call, together with the size N of the environment to be retained)
  - updates CP (next instruction);
  - loads Proc in P.

Note that trimming is performed only upon the execution of allocate, since it is the only block creation primitive. The call G,N instruction only transmits the N value, which will be used only if the execution of the goal G requires an environment allocation.

9.3.2.3. Example

Compiling the definition of qsort/3 in section 9.1.3, we get:

```
qsort/3 :    switch_on_term C1a, C1, C2, fail
C1a :        try_me_else C2a
C1 :         unify head against the Ai;
             proceed

C2a :        trust_me_else fail
C2 :         allocate
             unify head against the Ai;
             load the arguments of partition;
             call partition/4 , 6
             load the arguments of qsort;
             call qsort/3 , 3
             load the arguments of qsort;
             deallocate
             execute qsort/3
```

The fact is immediately translated through the unification of its arguments with the Ai registers and by the execution of proceed. For the rule, unification first is tried, then the goals in the body are successively called.

9.3.3. Unification

Unification is achieved between the head clause parameters and the arguments of a goal held by the Ai registers. Compilation benefits directly from the fact that we can:

1. statically determine the parameter type involved: constant, variable, list, or functional term;
2. in the case of a variable, distinguish its first occurrence (initial unbound state) from its other occurrences;
3. reason on the variables to determine whether they are held by the *local stack* or by the heap.

Consequently, the unification of each term type is processed by a (get) specialized instruction. In the case of a nonempty list or of a functional term each get instruction is completed by a unify instruction set that performs the subterms unification.

These unify instructions work in two modes (*read* and *write*). The first mode (*read*) allows the access to term instances (see section 5.1.5). The second mode (*write*) corresponds to constructing and forces a copying on the top of the heap.

9.3.3.1. *Get* Instructions

Depending on the argument type, and in the case of a variable whether or not it is its first occurrence, we use the relevant get instruction that performs unification between its two operands.

1. constant C: get_const C, Ai

2. variable Vn:

    a. first occurrence: get_variable Vn, Ai

    b. other occurrences: get_value Vn, Ai

3. list:

    a. empty list: get_nil Ai

    b. nonempty lists: get_list Ai

4. functional term: get_structure F, Ai

The get_list and get_structure instructions dynamically determine the working mode of the unify instructions that follow them. In both cases, if the dereferenced value of the Ai register leads to a structured term, the *read* mode is then chosen; otherwise, a new structure is allocated in the top of the heap, and the *write* mode is set for the copying of its arguments.

In the case of a variable standing as a parameter, the unification of its first occurrence is brought back to a simple binding through the get_variable instruction. However, get_value recursively unifies its two arguments on the PDL.

Let us consider the compiled form of the fact of the qsort/3 definition:

```
C1 :        get_nil A1              % []
            get_variable X1, A2     % R
            get_value X1, A3        % R
            proceed
```

A corresponding get instruction is associated with each parameter. The R variable is temporary and is thus numbered X1. When it is first used it is bound to A2, then it is unified with A3.

9.3.3.2. Unify Instruction

The unify instructions perform the unification of subterms. Here again, we distinguish three cases depending on whether the subterm is a constant, a variable, or a structured term:

1. C constant: unify_constant C

2. Vn variable:

    a. Vn's first occurrence: unify_variable Vn

    b. from the second occurrence and on:

    - unify_value Vn if we are positive that the first occurrence of Vn lead to a binding with a variable held on the heap,

    - unify_local_value Vn if we cannot draw any conclusion from the type of binding of Vn's first occurrence.

3. structured term: there are no specialized instructions for that case; we introduce temporary supplementary variables, and we generate the corresponding get and unify instructions.

The unify instructions work differently depending on whether they achieve constructing (*write* mode) or accessing (*read* mode). We thus again find the characteristics of the binding algorithm for structure copying (see section 5.1.5).

Let Arg be the argument on which we wish to perform unification. In the case of a C constant in *read* mode we unify it with Arg. In the *write* mode we copy C in the top of the heap.

For the first occurrence of a Vn variable, in *read* mode Vn is bound to Arg. In *write* mode, Vn is bound to a new unbound variable allocated on the top of the heap.

Beginning from the second occurrence we distinguish two cases depending on whether we were able to determine statically if the value of Vn was held by the heap. In *read* mode, unify_value unifies Vn and Arg using the PDL. In *write* mode we push the value of Vn on the heap.

The unify_local_value instruction performs the same task as unify_value, but tests first (mode *write*) if the value of Vn is a Y variable held on the *local stack*. If it is the case, a new unbound variable is pushed on the heap, and Y is then bound to that new variable.

Finally, in the case of a structured term, we introduce a supplementary temporary variable on which unification is achieved. This variable is then processed as if it were a "first level argument," that is, it is processed by a get instruction and the unify instructions that follow. This processing is repeated as many times as the depths of the involved subterms require it.

9.3.3.3. Example

Reconsidering the partial compilation of qsort/3 in section 9.3.2, and expanding the unification phase, we get:

```
qsort/3 :    switch_on_term C1a, C1, C2, fail
C1a :        try_me_else C2a
C1 :         get_nil A1                    % ([],
             get_variable X1, A2           % R,
             get_value X1, A3              % R)
             proceed
C2a :        trust_me_else fail
C2 :         allocate
             get_list A1                   % ([
             unify_variable Y6             % X |
             unify_variable X1             % L],
             get_variable Y5, A2           % R0,
             get_variable Y3, A3           % R)
             load the arguments of partition;
             call partition/4 , 6
             load the arguments of qsort;
             call qsort/3 , 3
             load the arguments of qsort;
             deallocate
             execute qsort/3
```

With respect to the rule, L is the only temporary variable (X1). The other six variables are permanent and numbered Y1 to Y6.

9.3.4. Loading the Ai Registers

As for unification, we differentiate among different cases depending on the type of the argument involved. We will first describe the case of a constant and that of a structured term and then cover the case of variables in an argument position. Finally, we will conclude by completing the compiled version of qsort/3.

9.3.4.1. Constant or Structured Term

Let us first consider the processing of constants and structured terms. Each case is managed by a particular instruction that loads the Ai register with the corresponding value:

1. Constant: put_const C, Ai

2. List:

   a. empty list: put_nil Ai

   b. nonempty lists: put_list Ai

3. functional term: put_structure F, Ai

The put instructions in the case of a nonempty list or a functional term are completed by the corresponding unify instructions. However, unlike the get primitives, all the unify instructions will be executed in write mode. Indeed, upon a call, every structured term argument is saved by a prior copying (see section 7.2.3).

9.3.4.2. Variable

The case of a variable in an argument position is more complex. Indeed, at its first occurrence it should be initialized as an unbound variable. Moreover, the unsafe variables should be dynamically reallocated on the heap in order to allow trimming (see section 9.1.3).

For a temporary variable Xn appearing as an argument, there are two possibilities:

1. On its first occurrence: put_variable Xn, Ai performs a dynamic reallocation of Xn on the top of the heap, as if it were an unsafe variable, then loads Ai.

2. From its second occurrence on: put_value Xn, Ai loads the Xn value into Ai;

In the case of a permanent variable Yn, two cases should be taken into account:

1. It is not the last goal where Yn occurs:

   a. it is Yn's first occurrence: put_variable Yn, Ai loads the Ai register while initializing Yn as unbound;

   b. From the second occurrence of Yn and on: put_value Yn, Ai loads Ai with the value of Yn.

2. It is the last goal where Yn occurs:

   a. Yn can then be an unsafe variable: put_unsafe_value Yn, Ai tests if Yn is unsafe, that is, initially unbound or bound to a variable from the same environment; if it does, then Yn is dynamically reallocated on the top of the heap, and Ai is loaded with its value; if it does not, then put_unsafe_value behaves like put_value;

   b. Yn cannot be a unsafe variable: then put_value Yn, Ai loads Ai with the Yn value.

The static determination of potentially unsafe variables is based upon the properties of their first occurrence. A self-init variable will never be unsafe. Indeed, if it satisfies $H2$ or $B2$ it will then be bound to a term allocated on the heap (see section 9.3.1). If an X variable satisfies $H1$, X will then be bound to the content of the corresponding Ai register. If it is a Y variable, then Y can only have a strictly lower creation time than that of the block holding X (see section 6.3.4).

9.3.4.3. Example

Reconsidering the partial compilation of qsort/3 in section 9.3.3, and expanding the loading of the registers for the rule, we get:

```
C2 :         allocate
             get_list A1              % ([
             unify_variable Y6        % X |
             unify_variable X1        % L],
             get_variable Y5, A2      % RO,
             get_variable Y3, A3      % R)

             put_value X1, A1         % (L,  occ. 2
             put_value Y6, A2         % X,   occ. 2
             put_variable Y4, A3      % L1, occ. 1
             put_variable Y1, A4      % L2) occ. 1
             call partition/4 , 6

             put_unsafe_value Y4, A1 % (L1,   unsafe ?
             put_value Y5, A2         % RO,    safe H1
             put_list A3              % [
             unify_value Y6           % X |    occ. 2
             unify_variable Y2        % R1])   occ. 1
             call qsort/3 , 3

             put_unsafe_value Y1, A1 % (L2,   unsafe ?
             put_value Y2, A2         % R1,    safe B2
             put_value Y3, A3         % R)     safe H1
             deallocate
             execute qsort/3
```

Before the execution of call qsort/3,3 the loading of Y6 is performed by unify_value and not by unify_local_value. Indeed, Y6 satisfies H2 and is allocated on the heap (see section 9.3.2). Y5 is self-init (H1) and cannot obviously be an unsafe variable.

For the last call to qsort/3, the Y2 and Y3 variables cannot be unsafe: Y2 is also self-init (B2), and for Y3 the situation is the same as for Y5.

Finally, let us observe that nothing can be statically determined for Y4 and Y1 since they have been both initialized by a put_variable instruction.

9.3.5. Temporary Variables Management

The temporary variables are directly managed through the A_i registers, without resorting to local space allocation during the duplication of the clause where they occur. We can map them directly to the corresponding A_i registers through an adequate numbering of the temporary variables; that is, Xi and Ai will point to the same memory location. Certain unification and loading steps then become immediate, the get_variable Xi, Ai and put_value Xi, Ai being rendered ineffective.

Let us review first the quicksort example and let us show that the code dealing with the unification to the first fact can be optimized through an adequate numbering of the R (X2) variable.

```
C1 :        get_nil A1              % ([],
            get_variable X2, A2     % R,
            get_value X2, A3        % R)
            proceed
```

Since A2 and X2 coincide, get_variable X2, A2 is rendered ineffective and we can replace the instruction that follows by get_value A2, A3:

```
C1 :        get_nil A1              % ([],
            get_value A2, A3        % R, R)
            proceed
```

By applying the same principle to the rule, we then obtain the final compiled form of qsort/3:

```
qsort/3 :   switch_on_term C1a, C1, C2, fail

C1a :       try_me_else C2a
C1 :        get_nil A1              % ([],
            get_value A2, A3        % R, R)
            proceed

C2a :       trust_me_else fail
C2 :        allocate
            get_list A1             % ([
            unify_variable Y6       % X |
            unify_variable A1       % L],
            get_variable Y5, A2     % R0,
            get_variable Y3, A3     % R)

            put_value Y6, A2        % (L, X,
            put_variable Y4, A3     % L1,
            put_variable Y1, A4     % L2)
            call partition/4 , 6
            put_unsafe_value Y4, A1 % (L1,
            put_value Y5, A2        % R0,
            put_list A3             % [
            unify_value Y6          % X |
            unify_variable Y2       % R1])
            call qsort/3 , 3
            put_unsafe_value Y1, A1 % (L2,
            put_value Y2, A2        % R1,
            put_value Y3, A3        % R)
```

```
                deallocate
                execute qsort/3
```

The numbering of temporary variables is based upon the variables that will end in an argument position, especially in the case of a rule with a single literal in its body. The compilation of conc/3 shows it significantly.

First of all, the compiled version of the fact and of the indexing are similar to qsort/3. For the rule, let us call X1, X2, and X3 the three variables that occur in the recursive call. Indeed, they are to be loaded in the A1, A2, and A3 registers respectively. We then obtain a first version:

```
conc/3 :     switch_on_term C1a, C1, C2, fail

C1a :        try_me_else C2a
C1  :        get_nil A1                % ([],
             get_value A2, A3          % R, R)
             proceed

C2a :        trust_me_else fail
C2  :        get_list A1               % ([
             unify_variable X4         % T |
             unify_variable X1         % Q],
             get_variable X2, A2       % L,
             get_list A3               % ([
             unify_value X4            % T |
             unify_variable X3         % R])
             put_value X1, A1          % (Q,
             put_value X2, A2          % L,
             put_value X3, A3          % R)
             execute conc/3
```

All the loading steps having been automatically performed, this leads us to the following final version:

```
conc/3 :     switch_on_term C1a, C1, C2, fail
C1a :        try_me_else C2a
C1  :        get_nil A1                % ([],
             get_value A2, A3          % R, R)
             proceed
C2a :        trust_me_else fail
C2  :        get_list A1               % ([
             unify_variable X4         % T |
             unify_variable A1         % Q],
             get_list A3               % L, ([
             unify_value X4            % T |
             unify_variable A3         % R])
             execute conc/3
```

9.4. Extensions

Numerous extensions and modifications of the WAM have been proposed, especially for the clauses indexing mechanism, the choice point management, and unification.

As observed in section 9.2.4, the management of clauses having a variable as their head's first parameter can lead to the creation of multiple choice points related to a single goal. Many fixes have been proposed: merging these choice points into a single one, the tradeoff being a new appropriate management of the BP field of the rerun parts [Tur86]; the use of one indexing level only [Car87]; or delaying as much as possible the creation of a choice point [BBC83].

Finally, many improvements have been made at the level of compiling unification and Ai registers management: the definition of specialized unify instructions for the *read* and the *write* modes, the reordering of the unification arguments in the *write* mode [GJ89], and the optimized allocation of structured terms [Mar88].

CHAPTER 10

THE *DIF* AND *FREEZE* PREDICATES OF PROLOG-II

The *freeze*/2 predicate of Prolog-II delays a goal as long as a variable remains unbound (see section 3.4.1). The *dif*/2 predicate enables the imposition of a constraint on two terms by forcing them to be different in the rest of the proof.

Even though $dif(t1, t2)$ is formally defined as being the addition of the inequality $t1 \neq t2$ to the current system of equations and inequations on rational trees [Col82b], its implementation uses the same delaying mechanism as the *freeze*/2 predicate (see section 3.4).

The main idea in the implementation of the delaying mechanism is to perform a direct association between the variable and the goal whose proof will be delayed [Can84]. This approach allows a direct access to the awakened goals, thus avoiding the search for the thawed goals among the set of the delayed goals after each unification.

However, in order to implement such a delaying mechanism, the following observations should be accounted for:

1. the triggering of a *dif*/2 is more sensitive than that of a *freeze*/2;

2. since a given variable can be associated with more than one delayed goal, a representation of these sequences of goals should be proposed;

3. the simultaneous triggering of multiple goals after a successful unification forces us to set a triggering strategy, that is, to specify in which order these different goals should be proved; and

4. the definition of a proof's success should take into account the delayed goals that will never be awakened.

Our strategy will consist in modifying the architecture and memory management proposed in chapter 7 according to the requirements of the delaying mechanism. Accordingly, the sequence of delayed goals will be stored in a dedicated memory area: the stack of frozen goals [Boi86a], [Boi86b]. The modifications that are necessary for a correct behavior at both the delaying level and the triggering one will then be made. Finally, we will demonstrate that last-call optimization is still applicable, despite these modifications.

The use of the *dif*/2 predicate will be described first, followed by the definition of the problem of implementing the delaying mechanism according to the Prolog-II solution [Can86]. An implementation in the context of structure sharing will then be proposed. Finally, our approach will be compared to that of SICStus Prolog [Car87], which uses structure copying.

10.1. The *dif*/2 Predicate of Prolog-II

The *dif*/2 predicate of Prolog-II allows for the constraining of two terms by forcing them to remain different in the rest of the proof. If necessary, dif subsequently continues to verify that any binding of a variable occurring in one of these two terms respects the initial inequality constraint.

The use of the *dif*/2 predicate will be illustrated through an example. A first implementation based on the delaying mechanism already developed for the *freeze*/2 predicate (see section 3.4) will then be described.

10.1.1. The Use of the *dif*/2 Predicate

The *dif*/2 predicate forces two terms $t1$ and $t2$ to remain different in the rest of the proof. When $dif(t1, t2)$ is called, if $t1$ and $t2$ cannot be unified, then $dif(t1, t2)$ definitely succeeds. If $t1$ and $t2$ are identical, $dif(t1, t2)$ fails. Otherwise, $dif(t1, t2)$ is delayed in order for the constraint to be perpetuated.

The use of *dif*/2 is an elegant means of solving numerous problems, which were until now difficult to transcribe in Prolog. The *dif*/2 predicate enables among other things the specification of constraints on structures that are initially unknown and to which we would like subsequently to attribute values. A common example is list construction, where all the elements should be different.

This processing is performed by the are_dif/1 predicate [Kan82]. Two cases are to be distinguished, depending on whether or not the first level of the list is known (predicates are_dif/1 and are_dif2/1 respectively)

```
are_dif([]).
are_dif([T|Q]) :- out_of(T,Q),
    are_dif(Q).

out_of(_,[]).
out_of(X,[T|Q]) :- dif(X,T),
    out_of(X,Q).

are_dif2(L) :- freeze(L,are_dif1(L)).

are_dif1([]).
are_dif1([T|Q]) :- out_of2(T,Q),
    are_dif2(Q).

out_of2(X,L) :- freeze(L,out_of1(X,L)).

out_of1(_,[]).
```

```
out_of1(X,[T|Q]) :- dif(X,T),
    out_of2(X,Q).
```

The first level of the list being instantiated, we traverse it by imposing the difference constraints two by two. However, if the first level is not determined, the `are_dif1/1` predicate needs to be delayed so that it can impose the constraints all through the construction.

Let us recall the encryption example send+more=money of M. Van Caneghem, in order to apply the `are_dif/1` predicate first version. We first specify that the list we are looking for is made up of unique digits. We are then left with the definition of addition, together with that of the carry process (predicate sum/5).

```
send :- sol_send([S,E,N,D,M,O,R,Y]),
    write_send([S,E,N,D,M,O,R,Y]).

send2 :- dif(M,O),
    sol_send([S,E,N,D,M,O,R,Y]),
    write_send([S,E,N,D,M,O,R,Y]).

sol_send([S,E,N,D,M,O,R,Y]) :- are_dif([S,E,N,D,M,O,R,Y]),
    sum(R1,O,O,M,O),
    sum(R2,S,M,O,R1),
    sum(R3,E,O,N,R2),
    sum(X,N,R,E,R3),
    sum(O,D,E,Y,X).

write_send([S,E,N,D,M,O,R,Y]) :- wl([' ',S,E,N,D]),
    wl(['+',M,O,R,E]),
    write('----------'), nl,
    wl([M,O,N,E,Y]).

sum(X,O,O,X,O) :- !, carry(X).
sum(R,X,Y,Z,R1) :- carry(R),
    digit(X),
    digit(Y),
    plus(X,Y,X1),
    plus(X1,R,T1),
    divi(T1,10,R1),
    mod(T1,10,Z).

digit(0).
.../...
digit(9).

carry(0).
carry(1).

wl([]) :- nl.
wl([X|Y]) :- write(X), write(' '), wl(Y).
```

The send/0 predicate produces twenty-five solutions whereas send2/0 has a unique one. For other examples on the use of the *dif*/2 predicate, the reader can refer to [GKPC86].

10.1.2. Implementation of *dif*/2

dif($t1, t2$) tries to unify the two terms $t1$ and $t2$. It then analyzes the effects of this unification:

1. if it fails, then *dif*($t1, t2$) definitely succeeds;
2. if it does not involve any binding of variables, then *dif*($t1, t2$) fails since the two terms are identical;
3. if it succeeds in binding at least one variable of $t1$ or $t2$, then the constraint should be kept since it is too early to diagnose success or failure.

Perpetuating the constraint consists in delaying it through a variable bound by the unification between $t1$ and $t2$ [Can84]. Indeed, the only case that corrupts the proof is that in which two terms become identical, for which backtracking should be applied. However, the fact that two terms become identical implies that this variable has been bound and that the delayed constraint has been activated.

10.2. Implementation Principles

Our description of the implementation of the delaying mechanism will be based on the Prolog-II solution [Can86]. The Prolog-II architecture will be described first, followed by the description of the main idea of the implementation that consists in directly binding the variable to the delayed goals. The problem of the representation of a sequence of delayed goals held by a unique variable will be described next (see section 10.2.3).

We will focus on the sensitivity of the triggering mechanism in section 10.2.4 where the Prolog-II approach (which consists in implementing *dif*/2 and *freeze*/2 through a more sensitive triggering mechanism, the *delay*/2 predicate) will be described. The triggering strategy will be covered in section 10.2.5: after observing that the *dif*($t1, t2$) goals should be proven before all others, we will discuss the problems inherent to the implementation of *freeze*/2 with *delay*/2. Section 10.2.6 will conclude by presenting the new concept of a proof success.

10.2.1. Prolog-II Architecture

Prolog-II [Can84],[Can86] is based on a structure sharing representation of terms. The working area of the interpreter is divided into five stacks:

1. the choices stack

2. the steps stack

3. the substitutions stack

4. the restoration stack

5. the equations stack.

Besides the equations stack (see section 1.4.2), which realizes the unification of rational trees, there are two basic differences between Prolog-II and Warren's Abstract Machine (see section 7.2.5).

First of all, the control is no longer managed by a single stack (i.e., the *local stack*): the choice points are now allocated on a special stack—the choices stack—whereas the forward process pushes every call of a new goal on the steps stack.

Moreover, the environment and control parts are completely separated. Each new environment created by the duplication of a clause is allocated on a special stack: the substitutions stack, which does not differentiate between local and global variables (see section 6.2.1). These choices, as we shall see, are made in order to facilitate the implementation of the freezing and triggering of the delayed goals.

10.2.2. The Predicate *freeze*/2

The main idea of this implementation is to associate the freezing variable directly with the delayed goal. Since the freezing of a goal can only occur if the variable is unbound, the variable's memory location is used to reference the delayed goal.

This approach results in two kinds of unbound variables: the authentic ones, which are unbound in the usual sense; and the frozen ones, which, although they are unbound from the point of view of unification, reference the delayed goals.

During a call of the form *freeze*$(X, Goal)$ two cases are to be considered, depending on the status of the variable X: X is either bound to a constant or a structured term (and *Goal* is immediately proven) or unbound (and *Goal* is delayed). If X is already frozen, then it is bound to a new sequence of goals composed of the existing one, xl, to which b is added.

freeze(X,Goal)
begin
        if X is bound to a constant or a structured term,
            then call Goal;
            else

        if X is truly unbound,

            then associate(X,Goal,empty-sequence);

            else

            let xl be the sequence of delayed goals associated to X;

            associate(X,Goal,xl);

        endif;

    endif;

end;

This direct association between freezing variables and delayed goals has the advantage of allowing an easy triggering of the frozen goals [Can84]. Indeed this approach avoids a systematic search among all delayed goals for the ones to be awakened. Upon each binding of a variable X with a constant or a structured term, it will be sufficient to check if X is bound to one or more delayed goals. If it is, a direct access to the goals to be awakened will then be provided.

10.2.3. Representation of the Delayed Goals Lists

Such an implementation of the delaying mechanism (i.e., by direct binding between variables and frozen goals) dynamically creates new sequences of goals: many delayed goals can be associated with a single variable at a given time (see the function *associate* in section 10.2.2).

Prolog-II represents such sequences of goals as lists created dynamically on the substitutions stack. Upon each addition of a new frozen goal *Goal*, two pairs are allocated on the top of this stack: the first one represents *Goal* and its environment, and the second is associated with the rest of the sequence. The Prolog-II substitutions stack thus represents both the environments and the sequences of delayed goals.

During the unification, the sequences of goals awakened through the binding of their freezing variables are gathered in order to be pushed on the top of the steps stack [Can84]. The proof then continues, thus triggering the execution of the goals that have just been awakened.

10.2.4. The Triggering Mechanism Sensitivity

In order to implement the *dif*/2 and *freeze*/2 predicates through the same delaying mechanism, we should conceive a more sensitive triggering mechanism than the one initially needed for the *freeze*/2 predicate alone. The problem will first be defined, after which the Prolog-II solution will be described. This solution consists in expressing *dif*/2 and *freeze*/2 through the *delay*/2 predicate, which has a more sensitive triggering mechanism.

10.2.4.1. The Problem

According to the semantics of the *freeze*/2 predicate, the triggering of the delayed goals should occur after each successful unification that has produced the binding of one or more frozen variables to a constant or a structured term. The *dif*/2 predicate, however, requires a more sensitive triggering strategy.

Consider the following example:

```
test :- dif(f(X), f(Y)), eq(X, Y).
```

The execution of test/0 should produce a failure since the two terms become identical after the call of eq(X,Y). It is thus necessary that the delayed constraint be awakened during the unification of X and Y, even though neither X nor Y is bound to a constant or a structured term.

The implementation of the *dif*/2 and *freeze*/2 predicates through the same delaying mechanism then forces a more sensitive triggering strategy than the one required by the *freeze*/2 predicate alone.

10.2.4.2. The delay/2 Predicate

The definition of *delay*(*X, Goal*) is the same as the one of *freeze*(*X, Goal*) (see section 10.2.2), except that *Goal* will be awakened when *X* will be bound to a constant, a variable, or a structured term.

Note that the semantics of *delay*/2 depends on the binding mechanism between unbound variables and on the numbering of the variables within a clause. This is illustrated by the following example:

```
test1 :- delay(X,write(1)), eq(X,Y).
```

If X is the variable numbered 0 and Y is the variable numbered 1, then the unification between X and Y will produce the binding between Y and X according to the binding mechanism between two unbound variables (see section 5.1.3), and the goal will still be delayed. On the other hand, if the numbering is reversed, X will then be bound to Y and write(1) will be awakened. Notice that this behavior of *delay*/2 will not affect the *dif*/2 and *freeze*/2 predicates.

10.2.4.3. *dif* and *freeze* in Terms of Delay

dif/2 and *freeze*/2 can then be expressed in terms of *delay*/2 [Can86]:

```
freeze(X, B) :- var(X), !,
    delay(X, freeze(X, B)).
freeze(_, B) :- call(B).

dif(T1, T2) :- special_id(T1, T2, V), !,
    var(V),
    delay(V, dif(T1, T2)).
dif(T1, T2).
```

Freeze is performed by perpetuating delaying as long as X is unbound. Although this implementation of `freeze/2` through `delay/2` is simple, it is nevertheless expensive for all the triggering/delaying caused by bindings between unbound variables.

The built-in predicate `special_id(T1,T2,V)` is defined as follows: it first tries to unify T1 and T2 and then analyzes the effects of this unification:

1. if it fails, then `special_id(T1,T2,V)` fails (the two terms are different);
2. if it succeeds with no variable binding, then `special_id(T1,T2,V)` succeeds by binding V to the atom `true` (the two terms are identical);
3. if it succeeds in binding at least one variable X of T1 or T2, then `special_id(T1,T2,V)` succeeds by binding V to X (it is too early to decide).

Consequently (see section 10.1.2), `dif(T1,T2)` succeeds through the second clause if the two terms are different. It fails in case the two terms are identical because of the first rule. Otherwise, the constraint is delayed on the X variable of T1 or T2, which will be bound in order to make T1 and T2 identical (the only potential failure of `dif(T1,T2)`).

10.2.5. Triggering Strategies

The simultaneous triggering of multiple goals after a successful unification forces the definition of a triggering strategy, that is, to specify in which order these different goals should be proved. Furthermore, because of the very nature of the *dif*/2 predicate (nonidentity constraint) the *dif* goals should first be proven. Finally, even if we choose an explicit triggering strategy, the implementation of *freeze*/2 through *delay*/2 can produce behaviors that are difficult for the user to control.

10.2.5.1. The Triggering Order

In the case of a simultaneous triggering of multiple goals, the next step is to decide in what order these goals should be activated. As [RH86] observes, two approaches are then feasible.

The first approach—undoubtedly the easiest one from the point of view of the implementor—is to reactivate the different sequences of goals according to the order in which the freezing variables have been bound; and for a sequence of goals itself, to trigger the goals that constitute it according to the implicit order induced by the binding mechanism. This approach is, however, difficult to accept from a user's point of view since he or she will not be able to control the simultaneous triggering of different goals without deeply analyzing the internal binding mechanism.

The second approach consists in triggering the goals according to their freezing time, that is, the first frozen goal is proved first (queue mechanism). This is the approach retained by MU-Prolog [Nai85a],[Nai85b] for the implementation of the *wait declarations* (see section 3.3).

A second solution consists in reactivating the goals according to the reverse order in which they have been delayed, that is, the last frozen goal is first proved (stack mechanism). This approach is the one chosen in IC-Prolog [CCG82] (see section 3.2).

10.2.5.2. The *dif(s)* First

Even though the *dif*/2 and *freeze*/2 predicates have similar implementations due to the use of the same delaying mechanism, their semantics are completely different [RH86]. *freeze*/2 is a control mechanism, whereas *dif*/2 is an inequality constraint on two terms.

The *dif* goals should always be activated first during the triggering stage, otherwise, premature thaws [RH86] will appear. Consider the two following definitions:

```
test1 :- dif(X,0),
    freeze(X,divi(100,X,R)),
    eq(X,0).

test2 :- freeze(X,divi(100,X,R)),
    dif(X,0),
    eq(X,0).
```

Regardless of which triggering strategy is chosen (queue or stack), test1 or test2 will lead to the premature thaw of the divi(100,X,R) goal, thus causing a serious malfunction. Indeed, this thaw leads to a division by zero, which is precisely what we were trying to avoid by specifying the dif(X,0) constraint.

Consequently, the *dif* goals should be proven first during the triggering phase. It is enough to consider their freezing time as being always prior (for a triggering strategy based on the delaying order) or subsequent (for a triggering strategy based on the reverse order) to that of all the other delayed goals.

10.2.5.3. Freezing Times: Freeze or Delay?

As observed above, the implementation of *freeze*/2 through *delay*/2 leads to triggering/delaying as bindings are performed between the frozen variables and unbound ones. Consequently, the freezing time of a goal will be that of the last delay that has been performed on it, and not automatically that of its initial freeze.

Let us consider the following example:

```
test :- freeze(V0,write(1)),  freeze(V0,write(2)),
    freeze(V1,write(3)),  freeze(V1,write(4)),
    freeze(V2,write(5)),  freeze(V2,write(6)),
    eq(V0,V2),
    eq(V2,V1),
    eq(V0,a).
```

Consider now a triggering strategy à la MU-Prolog. `test/0` should produce the `123456` printout. However, because of the `freeze/2` implementation through `delay/2`, `test/0` prints `125634`.

Indeed, `eq(V0,V2)` produces the binding between V2 and V0 and leads to the triggering of `freeze(V2,write(5))` and `freeze(V2,write(6))` immediately delayed on V0 again. `eq(V2,V1)` then binds V1 to V0, triggering `freeze(V1,write(3))` and `freeze(V1,write(4))` delayed again on V0.

Let us describe the binding status and the delayed goals throughout the execution of `test/0`.

```
{}        V0 -> freeze(V0,write(1)),  freeze(V0,write(2))
          V1 -> freeze(V1,write(3)),  freeze(V1,write(4))
          V2 -> freeze(V2,write(5)),  freeze(V2,write(6))

{V2=V0}   V0 -> freeze(V0,write(1)),  freeze(V0,write(2)),
                freeze(V0,write(5)),  freeze(V0,write(6))
          V1 -> freeze(V1,write(3)),  freeze(V1,write(4))

{V2=V0; V1=V0}
          V0 -> freeze(V0,write(1)),  freeze(V0,write(2)),
                freeze(V0,write(5)),  freeze(V0,write(6)),
                freeze(V0,write(3)),  freeze(V0,write(4))

{V2=V0; V1=V0; V0=a}
```

The queuing strategy is respected at each step, each new delayed goal being appended to the end of the set of goals already delayed. The control of the triggering mechanism by the user then becomes very delicate, since the freezing time taken into account for a goal is that of the last delay that has been performed on it.

10.2.6. The Concept of Success

Despite the introduction of the delaying mechanism, the proof has always been considered to end when all the goals to be proved have succeeded. Whether the goal is delayed or not, it should be considered as an intrinsic part of the resolvent. Since the proof ends when the resolvent is empty, therefore, the delayed goals that have never been awakened should be taken into account.

As observed by L. Naish [Nai85b], a proof can then terminate in three ways: a definitive failure, a success with an empty resolvent, or a success with nonawakened delayed goals.

The third case should then be dealt with separately from the second one, by associating with it a warning (for example) that will notify the user of the presence of still-unawakened frozen goals. The Prolog system could then be provided with a built-in predicate that would allow at a given time the access or the visualization of the different unawakened delayed goals.

The use of such a predicate is twofold: it enables the detection of the empty resolvent at the end of the proof by verifying the correctness of the proof, and it is also a debugging tool for the programs that rely heavily on the delaying mechanism.

10.3. An Implementation of *dif*/2 and *freeze*/2

An implementation of *dif*/2 and *freeze*/2 will be proposed in the context of structure sharing, which is the representation mode retained by Prolog-II. Our approach will consist of adapting the architecture and memory management proposed in chapter 7 to the requirements of the delaying mechanism.

Consequently, the sequences of delayed goals will be held in a dedicated memory area: the stack of frozen goals [Boi86a], [Boi86b]. A triggering strategy that takes into account the constraints defined in section 10.2.5 will be chosen: the $dif(s)$ first, followed by the other goals according to a queuing strategy à la MU-Prolog.

This implementation will allow a triggering of the goals according to their initial freezing time (see section 10.2.5), thus avoiding the problems relative to an implementation with the *delay*/2 predicate. Furthermore, the stack strategy will provide a simple and efficient way of detecting the unawakened goals and the corresponding freezing variables.

10.3.1. The Stack of Frozen Goals

The sequences of frozen goals are represented in a special memory area—the stack of frozen goals—updated only upon backtracking [Boi86a],[Boi86b].

Each block allocated on this stack has four fields:

1. $FGvar$: the frozen variable;

2. $FGtail$: the rest of the sequence of goals delayed on the same variable;

3. $FGgoal$: the delayed goal;

4. $FGenv$: the execution (global) environment of this goal.

The execution environment of a delayed goal is evidently global: any variable that appears in a delayed goal (structured term) will be first classified as global (see section 6.2.1). Moreover, the execution of a frozen goal will not result in any access to local environments since a global variable cannot be bound to a local one (see section 6.2.5).

Such blocks will be allocated during each delaying, either explicitly through a *dif* or *freeze* call (see section 10.3.2), or implicitly through the action of perpetuating a *freeze* during the triggering phase (see section 10.3.3).

10.3.2. The Implementation of *dif* and *freeze*

The binding mechanism between a variable and a sequence of delayed goals will be described first. The two built-in predicates *dif*/2 and *freeze*/2 will then be implemented.

10.3.2.1. Binding between a Variable and a Sequence of Delayed Goals

An unbound frozen variable X should remain unbound while still allowing the access to the sequence of frozen goals xl associated with it. In this respect, the pair representing X is fully exploited: the first element still designates a particular value representing the unbound state, whereas the second element points now to the frozen block representing xl.

The delaying of a new goal b is performed in two steps: a new block is allocated in the top of the stack of frozen goals, then the X variable is associated with that newly created block. This operation is performed by the $bindfg$ primitive.

Consider the initial condition (Figure 10.1) where the X variable is already bound to a sequence of delayed goals, accessible through its frozen block l. $bindfg(x, b, eb, l)$ first adds the goal b (that has the eb execution environment) to the sequence l of goals already delayed on x, thus creating a new sequence $l1$. $bindfg$ then performs the binding between x and the new sequence of goals hence created $l1$ (see Figure 10.2).

Accordingly, a new block is allocated on the top (FR) of the frozen goals stack. Its first field $(FGvar)$ will designate x, and its second field $(FGtail)$ will reference the frozen block l. Finally, the third and fourth fields $(FGgoal,FGenv)$ will hold the b goal in its execution environment eb. The binding is then performed by providing the second part of the pair representing x with a reference to the newly created frozen block $l1$.

10.3.2.2. Implementing the Freeze Predicate

freeze$(X, Goal)$ is implemented as described in section 10.2.2.

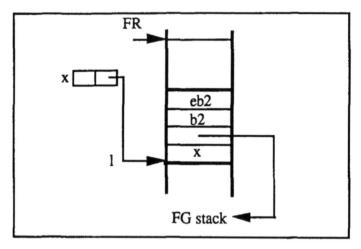

FIG. **10.1.** Binding between a variable and a sequence of delayed goals (initial condition)

freeze(X,*Goal*)
begin
      let xd be the dereferenced value of X;
      let b be the dereferenced value of *Goal*, an eb its environment;
      if xd is a constant or a structured term
            then call b;
            else
            if xd is a truly unbound variable
                  then $bindfg(xd, b, eb, empty - sequence)$;
                  else
                  let lg be the frozen block already bound to xd;
                  $bindfg(xd, b, eb, lg)$;
            endif;
      endif;
end;

The two association cases (see section 10.2.2) are dealt with through the *bindfg* primitive, depending on whether its call is performed with an empty sequence or an already existing one as its fourth argument.

10.3.2.3. Implementing the *dif* Predicate

$dif(t1, t2)$ tries first to unify the two terms $t1$ and $t2$ and then to analyze the results of this attempt (see section 10.1.2). However, since the effects of this unification should not remain, any variable binding is forced to be recorded on

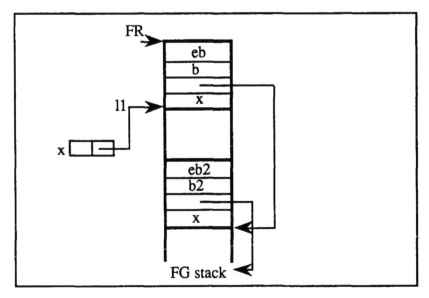

FIG. 10.2. A new sequence of goals

the *trail*. In this effect the backtracking management registers are temporarily updated with the current values of the tops of the stacks. All bindings will consequently be trailed. All that remains to do then is to restore the state of the bindings prior to the unification.

dif(t1,t2)
begin
    let e be the environment of $dif(t1, t2)$;
    assign BL and BG to the tops of the *local* and *global* stacks;
    let str be the current value of the top of the *trail*;
    let u be the result of $unify(t1, t2)$;
    restore BL and BG to their old values;
    if u is a failure
        then restore up to str;
        success;
        else
        if the current top of the *trail* = str
            then failure;
            else
        let x be the variable located at the top of the *trail*;
        restore up to str;
        let xl be: x's sequence of delayed goals if x is frozen,

else the empty-sequence;
$$bindfg(x, dif(t1, t2), e, xl);$$
   endif;
  endif;
end;

In order to determine the existence of at least one variable bound by the unification it is sufficient to compare the *trail*'s top values before and after unification. If they are equal, then $dif(t1, t2)$ produces a failure: the two terms are identical since their unification is performed with no variable binding. Otherwise, we choose the last variable that has been bound by unification, and we delay the constraint on it through the allocation of a new block on the stack of frozen goals.

10.3.3. Triggering Mechanism

The sensitivity of the triggering mechanism is the same as that of the *delay*/2 predicate (see section 10.2.4). Every time a frozen variable is bound (to a constant, another variable, or a structured term) the goals delayed on this variable are awakened.

Following a successful unification, the goals just awakened are proved in the following order: first the *dif(s)*, then the others (see section 10.2.5).

For the latter, we should distinguish between two cases: if their freezing variable is still unbound, then the goals remain delayed (premature triggering due to the binding between two unbound variables), or else they are proven according to their freezing time (i.e., according to a queuing strategy à la MU-Prolog [see section 10.2.5]).

In order to perpetuate the delaying without losing the initial freezing time (see section 10.2.5) a new frozen block is allocated whose $FGgoal$ field will no longer designate the delayed goal but the frozen block in which this goal appears. Thus, this indirection during the triggering phase will provide us with the real freezing time of the goal.

Given this information (effective time at which each goal has been delayed), it will be possible to perform the triggering while avoiding the problems relating to the implementation of the *delay*/2 predicate (see section 10.2.5).

10.3.4. The Concept of Success

This implementation of the delaying mechanism using the stack of frozen goals provides a simple and efficient means of detecting the still-unawakened goals together with the variables on which delaying is imposed.

In order to determine the unawakened delayed goals at a given time during the proof, it is sufficient to follow the stack of frozen goals sequentially by checking for each block if the variable V (*FGvar* field) is still in its initially unbound state. If it is, then the goal referenced by the block has not yet been awakened.

If the V variable is bound at that instant to a constant or a structured term, then it is certain that the goal has been awakened. If the V variable is bound to another variable X, then the goal is not to be considered as delayed on V since it is already delayed on X.

Let us recall the example of section 10.2.5 and include the visualization of the delayed goals still not awakened just prior to the call of eq(V0,a). This processing is performed by the built-in predicate frozen_goals/0:

```
| ?- listing(test).
   test :- freeze(V0,write(1)),  freeze(V0,write(2)),
        freeze(V1,write(3)),  freeze(V1,write(4)),
        freeze(V2,write(5)),  freeze(V2,write(6)),
        eq(V0,V2),
        eq(V2,V1),
        write('still frozen until now'), nl,
        frozen_goals,
        eq(V0,a).

| ?- test.
   still frozen until now
   write(3) frozen upon X3000
   write(4) frozen upon X3000
   write(5) frozen upon X3000
   write(6) frozen upon X3000
   write(2) frozen upon X3000
   write(1) frozen upon X3000
   123456
```

Besides the detection and correctness of the end of a proof (see section 10.2.6), the frozen_goals/0 predicate proves very useful in the debugging of programs that use the delaying mechanism.

10.3.5. Implementation

We will first describe the modifications that should be made to the architecture described in chapter 7 in order to implement the delaying mechanism. The properties of the *local stack* will be demonstrated as still valid. Consequently, the last-call optimization can then be implemented as described in chapter 7.

10.3.5.1. The Modifications

Besides the introduction of a fourth stack in the interpreter's working area (see section 10.3.1) two modifications appear to be necessary:

1. modification of the rerun part in order to be able to update the frozen goals stack upon backtracking,
2. modification of the *trail* management so that it takes into account the postbindings of the unbound frozen variables.

Foremost, the stack of frozen goals should be updated upon backtracking. Accordingly, each rerun part of a choice block will have a supplementary field (FR) indicating the top of this stack. A rerun part will from now on be composed of eight fields (see section 7.2.5): A, FR, BCP, BCL, BG, BL, BP, and TR. Examining the last choice point FR field during backtracking will allow the updating of the frozen goals stack.

Finally, the *trail* management should be modified so as to handle the unbound frozen variables that are postbound. Two informations (instead of one previously) will thus be saved on the *trail* during the postbinding of a frozen variable: the variable itself (i.e., its address) and the value of the second element of the pair that represents it (i.e., the associated frozen block).

10.3.5.2. Validation of the Memory Management Mechanisms

Despite the modifications just brought to the initial architecture proposed in chapter 7, the *local stack* memory management mechanisms—and especially the last-call optimization—remain valid. Indeed, the initial properties of the stacks are preserved.

First of all, the *local stack* retains its intrinsic properties: the distinction (see section 6.2.1) between local and global variables together with the respect of creation times during any L2-type binding (see section 10.2.2) insures the lack of bottom-up references on this stack.

Furthermore, the stack of frozen goals never accesses the *local stack* (see section 10.3.1): the fields of a frozen goal always refer to the *global stack* ($FGenv$), to the clauses coding zone, or to the frozen goals stack itself ($FGgoal$, $FGtail$). As for $FGvar$, it is sufficient to force any variable that appears in a *dif* or a *freeze* to be global. Consequently, this field will always refer to the *global stack*.

10.4. Comparison with SICSTUS Prolog

SICStus Prolog [Car87] proposes an implementation of the *dif*/2 and *freeze*/2 predicates in the context of structure copying. The Warren's Abstract Machine (see chapter 9) is extended to support both predicates. These are implemented through the *delay*/2 predicate (see section 10.2.4) whose triggering sensitivity is superior to that of *freeze*/2.

SICStus Prolog chooses to allocate the sequences of delayed goals on the heap as freezings are performed, and to keep records on the *trail* of all the bindings that occur between variables and delayed goals. The objective of this latter choice is to allow for the detection of unawakened goals at the end of a proof.

Delaying is performed classically, that is, by direct binding between the variable X and the goals whose proof has been delayed. A suspension is created on the heap upon each freezing of a goal b. Each suspension is a pair (new unbound variable, sb) where sb is either the copying of b if X is truly unbound, or the conjunction of b'copying with the sequence of goals already delayed on X. X is then bound to that suspension and this binding is automatically recorded on the *trail*.

Investigating the *trail* allows then the detection of unawakened goals at the end of the proof: among all the blocks on the *trail*, those that correspond to an unbound suspension (see section 10.3.4) designate the sequences of unawakened goals.

The essential difference between the two implementations is in SICStus Prolog's attempt to re-use the WAM memory areas, whereas we decided to implement a separate management: the frozen goals stack.

Aside from the term representation mode, the sequences of delayed goals are created in a similar fashion. Each goal term is a classical pair ($FGgoal$, $FGenv$) because of structure sharing, and $FGtail$ represents a sequence of goals (see section 10.3.1). For SICStus Prolog, copying and conjunction creation of goals are performed on the heap.

In our implementation, a binding is recorded on the frozen goals stack ($FGvar$), whereas for SICStus Prolog it is recorded directly on the *trail*. The variable copying upon each suspension allows a homogeneous processing of the *trail*: we are not obliged to account specially for unbound frozen variables (see section 10.3.5). On the other hand, the systematic trailing of all the suspensions can lead to useless restorings during backtracking.

Part III

Implementations

The goal of this third part, which is dedicated to implementation, is to employ the principles and implementation techniques described in Part 2. The sequence followed in Part 2 will be preserved in the construction of three different implementations (Mini-CProlog, Mini-WAM, and Mini-Prolog-II), related to the three levels previously described.

1. Mini-CProlog is a CProlog implementation (see section 6.2). It has the following characteristics: structure sharing, a three-stack memory management, and updating of the *local stack* upon return from deterministic calls.

2. Mini-WAM is based on the solution described in chapter 7. It can thus be considered as an interpreted implementation of Warren's Abstract Machine (see chapters 7 and 9). It has the following characteristics: structure copying, a working space made up of three stacks and the updating of the *local stack* by last-call optimization.

3. Mini-Prolog-II implements the *dif*/2 and *freeze*/2 predicates described in chapter 10. Its characteristics are structure sharing, a four-stack memory management, the updating of the *local stack* through last-call optimization, and the implementation of the delaying mechanism.

Finally, each of these three implementations will embed the clause indexing mechanism described in chapter 8.

A Unified Approach

Despite the different terms representations, memory managements, and control alternatives, a unified approach covering the three implementations will be undertaken by considering:

- an identical coding of the static structures,
- an identical representation of the working zone as a vector,
- a similar labeling of the working registers (borrowed from [War80]),
- similar function names for identical tasks,
- an identical management of the built-in predicates.

Each version will have its complete Common-Lisp code conveniently listed in the corresponding appendix. Since Mini-CProlog and Mini-Prolog-II have the same representation of terms, the common parts will be extracted and listed in appendix D.

Presentation Outline

Chapter 11 will be devoted to Mini-CProlog. The coding of static structures, common to the three implementations, will be presented first. The version elaborated in section 6.2 will be implemented next in the context of structure sharing. The management of the three stacks, unification, and the control implementation will then be elaborated respectively. At that point, the updating upon return from deterministic calls will be detailed.

Chapter 12 will describe Mini-WAM. First, unification in the context of structure copying will be described. Next, the same presentation as for Mini-CProlog will be followed and the modifications for structure copying and for the implementation of last-call optimization will be described.

Chapter 13 will deal with Mini-Prolog-II. It will follow the architecture of Mini-WAM but in the context of structure sharing. It will be modified to suit the delaying mechanism requirements associated with the $dif/2$ and $freeze/2$ predicates (see chapter 10). Accordingly, the four-stack architecture will first be specified, the frozen variables binding mechanism will be described next, and finally control will be implemented by focusing on the problem of the awakened goals.

Extensive references will be provided for each of these three implementations, pointing to the implementation principle(s): chapters 4, 5, and especially 6 (for Mini-CProlog); chapters 5, 7, and 9 (for Mini-WAM); and chapters 7 and 10 (for Mini-Prolog-II).

Chapter 14 will cover the built-in predicates. A reduced set of these will be chosen on purpose, for which an implementation will be proposed.

CHAPTER 11

MINI-CPROLOG

Mini-CProlog is a CProlog implementation and corresponds to the first level described in Part 2. Consequently, Mini-CProlog takes into consideration the proposals of chapter 6 and is issued from the formulated specifications and from the alternatives that have been chosen in chapters 4 and 5.

Mini-CProlog has the following characteristics: a terms representation by structure sharing, a three-stacks working zone architecture, and the updating of the *local stack* upon return from deterministic calls. Mini-CProlog will also benefit from the clauses indexing mechanism presented in chapter 8.

Mini-CProlog is based on the solution presented in section 2 of chapter 6, so it can be considered as an interpreted version of Prolog-Dec10. In order to simplify the implementation, Mini-Cprolog will not distinguish either the void variables or the temporary ones (see section 6.2.1). This enhancement will be proposed as an extension at the end of this chapter.

The principal aspects of the implementation of Mini-CProlog will be presented hereafter: the coding of the static structures, the working zone architecture, unification, and control. The complete Mini-CProlog code is divided into two parts: a part common to Mini-Prolog-II, listed in appendix D; and the part specific to Mini-CProlog, listed in appendix A.

11.1. The Coding of Static Structures

Mini-CProlog static structures coding fully exploit the Common-Lisp memory zones, whether for the representation of simple terms or for the coding of the structured terms and clauses. The translation from the external form to the internal representation is performed by the Mini-CProlog parser, whose code appears in both appendix A (for the part specific to Mini-CProlog) and appendix D (for the part common to Mini-Prolog-II).

We will describe the term coding first, then the clauses coding; later the indexing mechanism will be implemented.

11.1.1. Terms

The Mini-CProlog constants (positive integers and atoms) are considered as Lisp constants and are represented as such in the corresponding Lisp memory areas. As for C-Prolog, character strings are automatically translated by the parser into the list of ASCII codes that constitute the string.

Each Mini-CProlog variable is a Lisp dotted pair whose first element is a key (L for a local variable, G for a global one) and whose second element is its position number in the clause where it occurs. The numbering always starts at zero for all variable types (see section 6.2.1).

Structured terms are represented as Lisp lists. In order to simplify processing—and especially the unification algorithm—a similar coding has been chosen for Prolog lists and functional terms.

```
(defmacro functor (des) '(car ,des))
(defmacro arity (des) '(cadr ,des))
(defmacro des (te) '(car ,te))
(defmacro var? (v) '(and (consp ,v) (numberp (cdr ,v))))
(defmacro list? (x) '(eq (functor (des ,x)) ' \.))
```

Any Mini-CProlog structured term is hence represented in a Lisp list format with its first element describing the functional symbol. The rest of the list represents the arguments. Each functional symbol is coded as a pair (identifier arity), which in the Mini-CProlog particular case of a list is (. 2).

The f(a,g(b),C) functional term translates into ((f 3) a ((g 1) b) (L . 0)), supposing that C is the first local variable. The [a,b,c] list will become ((. 2) a ((. 2) b ((. 2) c ())))).

A straightforward representation of Mini-CProlog lists would have consisted in translating these into their equivalent Lisp syntax. However, since lists should be distinguished from functional terms at the coding level, marking the lists pairs would have become necessary. Incidently, this marking technique will be used in the next chapter for the implementation of structure copying. For the sake of homogeneity among the three versions, this common representation of structured terms has been adopted.

11.1.2. Clauses Coding

Each Mini-CProlog clause (either a fact or a rule) is represented as a list of literals, together with the necessary information for its duplication (see section 6.2.3), that is, the number of its local and global variables. Each literal is itself a list whose car is the predicate and whose cdr is the list of its arguments.

The first two clauses that define partition/4 (see section 6.2.1) will be respectively coded as such:

```
((1 . 0) (partition () (L .0 ) () ()) )

((2 . 3) (partition ((. 2) (G . 0) (G . 1))
              (L . 0)
```

```
           ((. 2) (G . 0) (G . 2))
           (L . 1))
       (le (G . 0) (L . 0))
       (! 2)
       (partition (G . 1) (L . 0) (G . 2) (L . 1)) )
```

In the case of a fact the list is reduced to two elements: the dotted pair (number of local variables, number of global variables) followed by the fact itself. In the case of a rule, the cddr of the list allows to access its body:

```
(defmacro nloc (c) '(caar ,c))
(defmacro nglob (c) '(cdar ,c))
(defmacro head (c) '(cadr ,c))
(defmacro tail (c) '(cddr ,c))
(defmacro pred (g) '(car ,g))
(defmacro largs (g) '(cdr ,g))
```

11.1.3. Clauses Indexing

The first indexing level only (as presented in section 8.1.1) will be implemented, together with a different treatment for the empty list and the nonempty ones. Five primary keys will be used in this effect (atom, def, fonct, empty, list) depending on the type of the term under consideration (constant, variable, functional term, empty list, nonempty list):

```
(defun nature (te)
  (cond
    ((var? te) 'def)
    ((null te) 'empty)
    ((atom te) 'atom)
    ((list? te) 'list)
    (t 'fonct)))
```

Every subset is a list of clauses held by the predicate P-list, under the indicator that is its associated key. On each call of a goal, the type of its first argument is determined in order to access the corresponding subset:

```
(defmacro def_of (g)
  '(get (pred ,g)
        (if (largs ,g)
            (nature (car (ultimate (car (largs ,g)) PCE PCG)))
            'def)))
```

For a goal with an arity greater than or equal to one, the type of its first argument is first determined and the corresponding subset is consequently accessed. If the goal has zero arity, then the whole set of clauses is retained.

In the case of a predicate, the clauses distribution is performed at the parsing level by the $ macro character:

```
(set-macro-character
  #\$
  #'(lambda (stream char)
      (let* ( (*standard-input* stream) (c (read_code_cl)))
        (add_cl (pred (head c)) c 'def)
        (if (largs (head c))
            (let ((b (nature (car (largs (head c))))))
              (if (eq b 'def)
                  (mapc
                    #' (lambda (x) (add_cl (pred (head c)) c x))
                    '(atom empty list fonct))
                  (add_cl (pred (head c)) c b)))))
      (values)))

(defun add_cl (pred c ind)
  (setf (get pred ind) (append (get pred ind) (list c))))
```

After parsing and coding, the clause should be appended to all the subsets where it should occur. It is first appended to the subset associated with def, and then, if its head has an arity greater than or equal to one, the type of its first argument is determined. If it is a variable, the clause must be appended to all the other subsets. Otherwise, it is sufficient to append it to the corresponding subset.

11.2. The Working Zone Architecture

Mini-CProlog working zone architecture is a single Lisp vector divided into three parts that correspond to three stacks: the *local stack*, the *global stack*, and the *trail*. The general architecture of the working zone will be first presented. The management of each of the three stacks will be detailed afterward. Their corresponding updating primitives will then be implemented.

11.2.1. General Architecture

The working zone consists of a single Lisp vector Mem divided in three parts corresponding to the *local stack, global stack*, and *trail* respectively. It is important to have the *global stack* located below the *local stack* in order to facilitate the L2 bindings (see section 6.2.5).

Each lower boundary of a zone is defined by a particular value that describes the size of each stack (see Figure 11.1).

Besides the three stacks, some registers are also necessary (their names are borrowed from [War80]):

- TR: top of *trail*.
- L: top of the *local stack*.

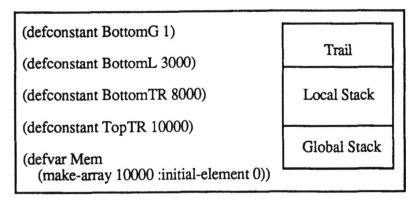

FIG. 11.1. The working area

- G: top of the *global stack*.
- CP: continuation point (i.e., the goal and its attached siblings).
- CL: local CP block (i.e., the father block in the context of the *and/or* tree).
- BL: local block for the last choice point.
- BG: global environment for the last choice point.
- PC, PCE, PCG: current goal and its two environments.

Each top of stack register indicates the first free location.

11.2.2. The *Local Stack*

The rerun part of the choice blocks will be described after presenting the deterministic blocks.

11.2.2.1. The Deterministic Blocks

Each deterministic block has four fields (CL, CP, G, E) that constitute the continuation and environment parts (see section 6.2.4). b designates the address of the beginning of the block for the four access primitives (see Figure 11.2). The push_cont primitive pushes the current continuation (CP, CL) on the *local stack*. L indicates the first free location.

The E field indicates the beginning of the local environment. This environment is made up of nl consecutive locations representing nl local variables. Since a structure sharing terms representation has been chosen, each location is a pair (see section 5.2.2)

An environment of size nl is allocated on the top of the *local stack* during the duplication of a clause that carries nl local variables (all the variables are initially unbound):

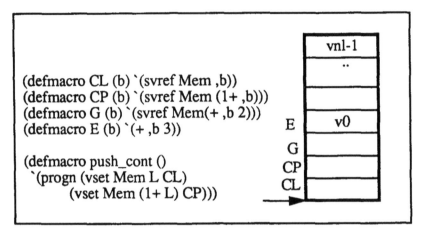

FIG. 11.2. Deterministic blocks

```
(defmacro push_E (n)
  '(let ((top (+ L 3 ,n)))
     (if (>= top BottomTR)
         (throw 'debord (print "Local Stack Overflow")))
     (dotimes (i ,n top)
       (vset Mem (decf top) (cons 'LIBRE BottomG))))))

(defmacro vset (v i x)
  '(setf (svref ,v ,i) ,x))

(defmacro maj_L (nl)
  '(incf L (+ 3 ,nl)))
```

The push_E primitive creates a local environment by initializing all the dotted pairs to the particular LIBRE atom (see section 5.2.1). In case of a *local stack* overflow, the current proof level is exited and the top-level loop is entered. The maj_L primitive terminates a local block by updating the top of the stack taking into account the size of the environment.

11.2.2.2. The Choice Blocks

Each choice block also consists of a rerun part represented by the three fields TR, BP, and BL (see section 4.2.4). In order to keep the same primitives that access the fields of the continuation and environment part, the address of a choice block will also be that of the CL field.

Now it is important to define the primitives for the allocation, modification, and deletion of choice points. The push_choix primitive allocates a rerun part. Accordingly the *trail* status (TR) and last choice point (BL) are pushed. The BP field will be subsequently updated by the push_bpr primitive. The block being

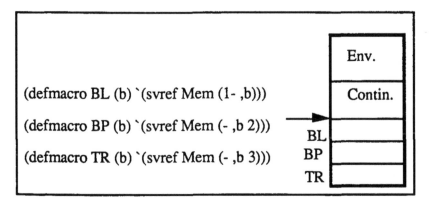

FIG. 11.3. Choice blocks

constructed becomes the current choice point (BL is updated) and BG points to the current top of the *global stack*:

```
(defmacro push_choix ()
  '(progn (vset Mem L TR) (vset Mem (incf L 2) BL)
          (setq BL (incf L) BG G)))

(defmacro push_bpr (reste)
  '(vset Mem (- BL 2) ,reste))

(defmacro pop_choix ()
  '(setq L (- L 3) BL (BL BL) BG (if (zerop BL) 0 (G BL))))
```

The push_bpr primitive updates the BP field of a choice block by associating it with the new set of clauses still to be considered. Finally, pop_choix deletes a choice point by updating L and by restoring BL and BG to the values that describe the previous choice point.

11.2.3. The *Global Stack*

An environment of a size ng is allocated on top of the *global stack* at each duplication of a clause carrying ng global variables. This environment has a structure similar to that of a local environment since it has ng locations corresponding to the ng global variables.

```
(defmacro push_G (n)
  '(let ((top (+ G ,n)))
     (if (>= top BottomL)
         (throw 'debord (print "Global Stack Overflow")))
     (dotimes (i ,n (vset Mem (+ L 2) G))
       (vset Mem (decf top) (cons 'LIBRE BottomG)))))

(defmacro maj_G (ng) '(incf G ,ng))
```

The push_G primitive operates like the push_E primitive, but on the *global stack*. It terminates by updating the G field of the local block under construction (see section 6.2.4). Here again, an overflow results in exiting the current proof and in reentering the top-level loop. The maj_G primitive terminates the global block by moving the G pointer to the top of the *global stack*.

11.2.4. The *Trail*

The *trail* detects the postbound variables in order to restore them to their initial values upon backtracking (see section 4.3.2). It is updated by two primitives:

```
(defmacro pushtrail (x)
  '(progn
     (if (>= TR TopTR) (throw 'debord (print "Trail Overflow")))
     (vset Mem TR ,x)
     (incf TR)))

(defmacro poptrail (top)
  '(do () ((= TR ,top))
     (vset Mem (svref Mem (decf TR)) (cons 'LIBRE BottomG))))
```

The pushtrail primitive pushes the variable that has the x address, whereas poptrail updates the *trail* up to the top address while restoring the variables encountered to their unbound initial state.

11.3. Unification

The primitives that access the environments and those that dereference the binding chains will first be described. The unification algorithm will then be presented. The binding mechanism implementation will conclude this section.

11.3.1. Primitives

An environment e is in fact the base address of the vector that carries the substitution. The concept of a term value within an environment should first be defined. In the case of a variable, the environment should be accessed. Otherwise (i.e., in the case of a constant or a structured term), the value is the term itself.

```
(defmacro adr (v e) '(+ (cdr ,v) ,e))
(defmacro value (v e) '(svref Mem (adr ,v ,e)))

(defun ult (v e)
  (let ((te (value v e)))
    (cond
      ((eq (car te) 'LIBRE) (cons v e))
      ((var? (car te)) (ult (car te) (cdr te)))
      ( te))))
```

```
(defun val (x e) (if (var? x) (ult x e) (cons x e))))
(defun ultimate (x el eg)
  (if (var? x)
      (if (eq (car x) 'L) (ult x el) (ult x eg))
      (cons x eg))))
```

The `ult` function dereferences a given variable (see section 5.1.3), the unbound status of a variable being implemented by a binding to the special atom LIBRE. The `ultimate` function performs the same processing as the function `val`, by first selecting the part of the environment in which the term is to be considered. A structured term is always picked up in the global environment. In the case of a constant, this does not matter. Finally, in the case of a variable, it depends on its type.

11.3.2. Unification Algorithm

The unification between a goal and the head of a clause consists in successively unifying each of their arguments:

```
(defun unify (t1 el1 eg1 t2 el2 eg2)
  (catch 'impossible
    (do () ((null t1))
      (unif (ultimate (pop t1) el1 eg1)
            (ultimate (pop t2) el2 eg2)))))
```

The `catch-throw` construction enables escape as soon as possible in case of failure. The `unif` function performs the unification between two terms in a dotted pair format:

```
(defun unif (t1 t2)
  (let ((x (car t1)) (ex (cdr t1)) (y (car t2)) (ey (cdr t2)))
    (cond
      ((var? y)
       (if (var? x)
           (if (= (adr x ex) (adr y ey)) t (bindv y ey x ex))
           (bindte (adr y ey) x ex)))
      ((var? x) (bindte (adr x ex) y ey))
      ((and (atom x) (atom y))
       (if (eql x y) t (throw 'impossible 'fail)))
      ((or (atom x) (atom y)) (throw 'impossible 'fail))
      ( (let ((dx (pop x)) (dy (pop y)))
          (if (and (eq (functor dx) (functor dy))
                   (= (arity dx) (arity dy)))
              (do () ((null x))
                (unif (val (pop x) ex) (val (pop y) ey)))
              (throw 'impossible 'fail)))))))
```

The case where one of the two terms is a variable is considered first. For two distinct variables, the `bindv` function then determines the binding direction before achieving it. Two different constants lead to a failure.

The case of two structured terms is considered last: after comparing their functional symbols and their respective arities, each argument is successively unified.

11.3.3. Binding Mechanism

The bindv primitive manages the case of a binding between two unbound variables. The bindte primitive performs the binding between the variable x and the term t represented by the pair (model, environment).

```
(defmacro bindv (x ex y ey)
  '(let ((ax (adr ,x ,ex)) (ay (adr ,y ,ey)))
     (if (< ax ay) (bindte ay ,x ,ex) (bindte ax ,y ,ey))))

(defmacro bindte (x sq e)
  '(progn
     (if (or (and (> ,x BottomL)(< ,x BL)) (< ,x BG))
         (pushtrail ,x))
     (rplaca (svref Mem ,x) ,sq)
     (rplacd (svref Mem ,x) ,e)))
```

bindv tests the creation times of the two variables in order to respect their binding direction (see section 5.1.3). Bindings always occur top to bottom by comparing addresses. Since the *global stack* is located beneath the *local stack*, this test works on both the case of two local variables and that of a local and a global one (see section 6.2.5).

The binding is performed by the bindte function that assigns the pair representing x with the (sq . e) term. The variable should be saved on the *trail* if it is older than the last choice point (see section 4.3.2). A variable is older than the last choice point if it is local and located beneath BL or if it is global and located beneath BG (see section 6.2.5).

11.4. Control

Prolog's control will be achieved by successively implementing the forward process and the backtracking. The updating upon return from deterministic calls will be detailed. The conditions of its application and its effective implementation will then be described respectively. The definition of the top-level loop will conclude this section.

11.4.1. Forward Process

The forward process selects the current goal (the left-most one) and executes it. Upon the call of the forward function the CP and CL registers respectively represent the current list of goals and the *local block* associated with it.

When the current proof is over (CP is empty) a first solution is obtained whose result is printed by calling the function answer. Otherwise, the first goal is selected by loading the PC, PCE, and PCG registers using the current continuation (CP,CL).

If it is a user goal, its execution starts with the set of candidate clauses (through the call of the pr function). If it is a built-in predicate, it is executed, and, depending on its success or failure, forward process is resumed or backtracking occurs. If the predicate is not defined then a failure is to be expected.

```
(defun forward ()
  (do () ((null Duboulot) (format t "no More ~%"))
    (cond
      ((null CP) (answer))
      ((load_PC)
       (cond
         ((user? PC)
          (let ((d (def_of PC))) (if d (pr d) (backtrack))))
         ((builtin? PC)
          (if (eq (apply (car PC) (cdr PC)) 'fail)
              (backtrack)
              (cont_det)))
         ((backtrack)))))))

(defun load_PC ()
  (setq PC (car CP) PCE (E CL) PCG (G CL)))

(defmacro user? (g) '(get (pred ,g) 'def))
(defmacro builtin? (g) '(get (pred ,g) 'evaluable))
```

Depending on the size of the set, the proof may or may not be deterministic. In the later case, a rerun part is pushed on the *local stack*:

```
(defun pr (paq)
  (if (cdr paq)
      (progn (push_choix) (pr_choice paq))
      (pr_det (car paq))))
```

The proof in the deterministic case will first be presented together with the updating of the *local stack*. The execution in the nondeterministic case will be described next.

11.4.1.1. Deterministic Case

A clause is duplicated (function pr_det) by allocating its local and global environments (see section 6.2.3). Unification is then performed between the current goal and the head of the clause. A failure leads to backtrack. If unification succeeds, the proof should continue by updating the CP and CL registers.

```
(defun pr_det (c)
  (if (eq (unify (largs PC)
                 PCE
                 PCG
                 (largs (head c))
                 (push_E (nloc c))
                 (push_G (nglob c)))
          'fail)
      (backtrack)
      (progn
        (if (tail c)
            (progn (push_cont)
                   (setq CP (tail c) CL L)
                   (maj_L (nloc c)))
            (cont_det))
        (maj_G (nglob c)))))
```

In this event, two cases are to be distinguished:

1. c is a fact, and it is not necessary to terminate the block, and L retains the value it had prior to the call (this is the particular case of a deterministic procedure that returns directly upon its call); the next goal to be proved is then fetched (call of the cont_det function);

2. c is a rule, thus the continuation should be preserved and the local block should be terminated; the new continuation is then obvious: it is the body of the rule.

In both cases, the global block should be terminated.

11.4.1.2. *Local Stack* Updating

The current goal having been proved by a fact, the continuation registers CP and CL should be updated. Two cases could then appear:

1. the goal was not in the last position. CP gets the rest of the list of current goals (sibling literals) as its value and CL remains unchanged (father block);

2. the goal was in the last position. In this case the call of the father goal is completely executed; if no intermediate choice point creation has occurred then the *local stack* can be updated. Notice that this procedure should be applied while the corresponding father goal is itself in the last position.

In order to determine the presence of intermediate choice points, the respective creation times of the last choice point (BL register) and the father block (register CL) must be compared.

The cont_det function deals with both cases:

```
(defun cont_det ()
  (if (cdr CP)
      (setq CP (cdr CP))
      (progn
        (if (< BL CL) (setq L CL))
        (do ()
            ((or (cdr (CP CL)) (zerop (CL CL)))
             (setq CP (cdr (CP CL)) CL (CL CL)) )
          (setq CL (CL CL))
          (if (< BL CL) (setq L CL))))))
```

If the call was in the last position, then the continuation parts are traced in order to fetch the next goal to be solved, while updating—if possible—the *local stack*. In this effect, if there are no intermediate choice points (BL<CL), the top of the *local stack* is reset to the address of the beginning of the father block (L:=CL). This process ends when a still-unresolved goal has been found (the rest of the CP field of CL is not empty), or when all the goals have been proved (i.e., there is no more father block).

11.4.1.3. Choice Case

In the case where a set contains many plausible clauses, the first one whose head satisfies unification is sought:

```
(defun shallow_backtrack (paq)
  (if (and (cdr paq)
           (eq (unify (largs PC)
                      PCE
                      PCG
                      (largs (head (car paq)))
                      (push_E (nloc (car paq)))
                      (push_G (nglob (car paq))))
               'fail))
      (progn (poptrail (TR BL))
             (shallow_backtrack (cdr paq)))
    paq))
```

Attempts are repeated as long as the set has multiple clauses and unification with the head of the first clause fails, while still restoring the binding status (call to poptrail). shallow_backtrack returns the set of clauses as soon as it is reduced to a single one.

If unification succeeds, shallow_backtrack returns the set whose first element is the clause that produced this success. Note that the clause is duplicated on each new unification by the allocation of the local and global parts of its environment.

In the nondeterministic case the proof is performed by the pr_choice function.

```
(defun pr_choice (paq)
  (let* ((re (shallow_backtrack paq)) (c(car re)) (r(cdr re)))
    (cond ((null r) (pop_choix) (pr_det c))
          ((push_bpr r)
           (push_cont)
           (if (tail c)
               (setq CP (tail c) CL L)
               (if (cdr CP)
                   (setq CP (cdr CP))
                   (do ()
                       ((or (cdr (CP CL)) (zerop (CL CL)))
                        (setq CP (cdr (CP CL)) CL (CL CL)))
                     (setq CL (CL CL)))))
           (maj_L (nloc c))
           (maj_G (nglob c)))))))
```

pr_choice monitors the work of shallow_backtrack. If the result has only one clause, we are back to the deterministic case. The rerun part is then popped and pr_det is activated. Otherwise, the first clause is that whose head satisfies unification and the rest of the set represents the BP field of the choice point. The block should then be terminated and the continuation registers should be updated by determining the next sequence of goals to be proven.

The activation block is terminated by updating the BP field and by pushing the continuation part.

The next sequence of goals to prove is then sought, considering the same cases as for a deterministic proof. If the clause is a rule, the continuation is then made up of the body of the rule and of the block L just allocated. Otherwise, two cases are to be differentiated, depending on whether the goal is in last position.

If the goal is not the last one to be considered, the continuation is then made up of the rest of CP and of CL, which remains unchanged. If the call is indeed the last one to be considered, the continuation fields should be examined in order to find the next sequence of goals to be proven. Unlike the deterministic case, no updating is attempted on the *local stack* since the last choice point is located on the top of the stack.

Finally, the two tops of the stacks (local and global) are updated depending on the size of the associated environments.

11.4.2. Backtracking

The backtrack function determines the presence of a choice point. If no choice point exists, then the proof is over. Otherwise, the three stacks are updated and the environments are restored. The PC, PCE, and PCG registers are loaded again and the forward process is restarted through the call of pr_choice, which has the set of clauses yet to be considered as its argument:

```
(defun backtrack ()
  (if (zerop BL)
      (setq Duboulot nil)
      (progn
        (setq L BL G BG
              PC (car (CP L))
              PCE (E (CL L)) PCG (G (CL L))
              CP (CP L) CL (CL L))
        (poptrail (TR BL))
        (pr_choice (BP L)))))
```

The goal to be proved again, together with the environment (local and global parts) in which it should be considered, is held by the continuation part of BL (see section 6.2.4).

11.4.3. The Top Level Loop

The top level loop initializes the different registers, creates the first activation block associated with the query, and then starts the forward process. Once all the proofs are performed, the same process is reactivated by parsing and encoding a new query:

```
(defun myloop (c)
  (setq G BottomG L BottomL TR BottomTR
        CP nil CL 0  BL 0 BG BottomG Duboulot t)
  (push_cont)
  (push_E (nloc c))
  (push_G (nglob c))
  (setq CP (cdr c) CL L)
  (maj_L (nloc c))
  (maj_G (nglob c))
  (catch 'debord (forward))
  (myloop (read_prompt)))
```

The top of each stack is initialized to the lower boundary of its respective zone. The initial continuation part is empty, and no choice point has yet been created. The activation block associated with the query is then allocated and the continuation registers are updated.

11.5. Extensions

Let us conclude by formulating some ideas for improvements:

  1. Introduce the concept of void and temporary variables. In this respect, the parser should be modified in order to differentiate the concepts of temporary and void variables. The allocation of the local part of the environment should then be modified together with the updating of the top of the stack in order to manage temporary variables (see section 6.2.3). Finally, the

definition of `unif` should be modified so that the unification "ignores" the void variables.

2. Implement the second level indexing as defined in section 8.1.

3. Propose a "convergent growth" of the *local* and *global stacks*.

   Instead of considering them as separate entities, we can force them to converge: the *local stack* would have a top-down allocation and the management of the *global stack* would remain unchanged.

   Indeed, this more flexible management will prove more suitable for the kinds of programs that consume one of the two resources in particular.

CHAPTER 12

MINI-WAM

Mini-WAM is a Prolog interpreter that corresponds to the second level, as described in Part 2. Accordingly, it implements the proposals of chapter 7. The Mini-WAM characteristics are as follows: a terms representation by structure copying, a three-stacks working zone architecture, and a *local stack* updating based on last-call optimization.

Mini-WAM corresponds to the solution proposed by D. H. Warren in 1983 (see chapter 7) and can thus be considered as an interpreted version of Warren's Abstract Machine (see chapter 9 for the specific compiling considerations). In order to simplify the implementation of Mini-WAM, neither trimming (see section 9.1.3) nor the differentiation between permanent and temporary variables (see section 9.3.1) will be implemented. These two enhancements will be proposed as an extension at the end of this chapter. Finally, Mini-WAM will benefit from clauses indexing as described in chapter 8.

With respect to Mini-CProlog, this second implementation brings two new features: first, structure copying replaces structure sharing; next, the last-call optimization replaces the updating upon return from deterministic calls on the memory management issue.

This presentation will try to refer as much as possible to Mini-CProlog. However, in order to emphasize the consequences of the new features, they will be explained separately. Accordingly, the modifications pertaining to the terms representation by structure copying will first be described. The working zone architecture will be studied next. The unification algorithm will then subsequently be implemented, and then the implementation of the control mechanism will be described, within which the last-call optimization will be detailed. The complete Common-Lisp code of Mini-WAM is listed in appendix B.

12.1. Structure Copying

Using structure copying forces a new representation of clauses and a different management of the *global stack*. Indeed, the difference between local and global variables no longer exists. Furthermore, the *global stack* allocation is performed dynamically, depending on the behavior of the copying algorithm.

A Lisp representation of the copy stack will be proposed after describing the new coding of clauses. The management primitives relative to this copy stack will then be defined, and the copying algorithm implemented.

12.1.1. Static Structures Coding

The terms models representation (simple or structured) of Mini-CProlog (see section 11.1.1) will be considered. A new clause coding should be provided since the difference between local and global variables inherent to structure sharing has vanished. The n variables of a clause will thus be numbered 0 to $n - 1$ (see section 6.3.2). A variable will be represented by a pair of the type of a V atom followed by its number in the clause where it appears.

Considering the example of section 11.1.2 we get:

```
( 1 (partition () (V.0) () ()))
( 5 (partition ((.2) (V.0) (V.1))
        (V.2)
        ((.2) (V.0) (V.3))
        (V.4))
    (1e (V.0) (V.2))
    (! 5)
    (partition (V.1) (V.2) (V.3) (V.4)))
```

In the case of a fact, the list is reduced to two elements: the number of variables, followed by the fact itself. In the case of a rule, the cddr of the list represents the body of the rule. The clause indexing mechanism of Mini-CProlog remains valid for Mini-WAM (see section 11.1.3): since indexing involves only static coding, it is independent from the representation of term instances.

12.1.2. Copy Stack Management

The Common-Lisp *List* zone has been chosen for the representation of the structured terms. This approach allows for a unique coding of the terms instances and models. The implementation of the unification algorithm and the binding algorithm will thus be simplified (see section 12.3). Furthermore, to duplicate a Common-Lisp vector element by element would have required the management of a new type of pointers (integers) and would thus have forced the choice of a different representation for integers. This solution has been rejected since it contradicts our initial specification of a basic unity between the three implementations.

However, since a stack management simulation is still required, each new instance thus created will be copied on the *global stack* in the form of a pointer to the *List* Lisp zone where the copy is located:

```
(defmacro recopy (x e)
  '(push_G (copy ,x ,e)))

(defmacro push_G (x)
  '(if (>= (incf G) BottomL)
```

```
(throw 'debord (print "Heap Overflow"))
(vset Mem G ,x)))
```

Taking into account this choice, the allocation on the *global stack* is no longer proportional to the size of the copied terms models and depends solely on the number of copies performed.

In order to differentiate the models from the structured terms instances (both of which are represented as lists), the latter will be marked. This is the same solution as the one retained in section 5.3.2, during the construction of a first unification algorithm that uses structure copying.

To copy a structured term model x in a given environment e consists then in copying the list representing x and marking it:

```
(defun copy (x e)
  (cond
    ((var? x)
     (let ((te (ult (adr x e))))
       (if (var? te) (genvar (cdr te)) te)))
    ((atom x) x)
    ((tcons (des x) (cop (largs x) e)))))

(defun cop (l e)
  (if l (cons (copy (car l) e) (cop (cdr l) e))))

(defun genvar (x)
  (bind x (push_G (cons 'V G))))

(defmacro copy? (te) '(cddr (des ,te)))

(defmacro tcons (des largs)
  '(cons (append ,des (list '#\*)) ,largs))
```

The problem of unbound variables will be found again in the copying of a term model (see section 5.3.1). It is resolved (see genvar) by the allocation of a new unbound variable on the top of the copy stack, and the binding (calling bind) of the initial variable to the newly created one. Unlike Mini-CProlog, the unbound state of a variable will no longer be symbolized by a particular atom but rather by the fact that the variable points to itself.

12.2. Working Zone Architecture

We will first describe the general architecture of the working zone while providing the Lisp representation of the saving registers. We will then present the *local stack* management derived from the blocks architecture described in chapter 7. Finally, the *trail* will be implemented.

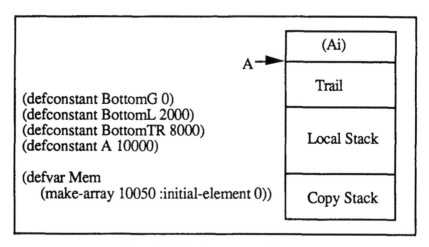

FIG. 12.1. Working area organization

12.2.1. General Architecture

The Ai registers will be represented as a vector held by the working zone. Accessing argument i is then performed through a simple offset from the base address A. The working zone is thus divided into four parts: the three stacks and the necessary space for the Ai registers (Figure 12.1).

The working registers are the same as those of Mini-CProlog (see section 11.2.1). Their definition, however, is that described in chapter 7:

- TR: top of *trail*,
- L: top of *local stack*,
- G: top of copy stack,
- CP: continuation point (next sequence of goals to be proved),
- CL: CP local block,
- BL: last choice block,
- BG: top of copy stack with respect to the last choice block,
- PC, PCE: current goal and its local environment.

Each top of stack register will always indicate the first free location, except G, which will designate the last occupied location, for the sake of convenience. Thus, the initial value BottomG will be initialized to 0, not 1 as in Mini-CProlog.

12.2.2. The *Local Stack*

The case of a deterministic call will first be considered. The choice points management will be described next.

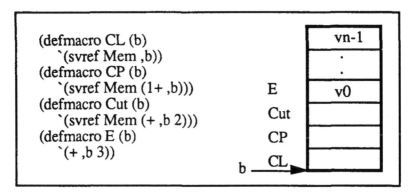

FIG. 12.2. Deterministic blocks

12.2.2.1. The Deterministic Case

Each deterministic block will be divided into its continuation and environment parts (see section 7.2.5). For all four primitives, b represents the base address of the block (Figure 12.2).

A supplementary field (Cut) has been added, which memorizes the last choice point preceding the call. Its unique purpose is to ease the implementation of the cut/0 predicate. More on this will follow in section 14.3.1, which deals with the built-in predicates.

Each time a clause holding n variables is duplicated, an environment of size n is allocated on the top of the *local stack*:

```
(defmacro push_E (n)
  '(let ((top (+ L 3 ,n)))
     (if (>= top BottomTR)
         (throw 'debord (print "Local Stack Overflow")))
     (vset Mem (+ L 2) Cut_pt)
  (dotimes (i ,n top) (vset Mem (decf top) (cons 'V top)))))

(defmacro push_cont ()
  '(progn (vset Mem L CL) (vset Mem (1+ L) CP)))
(defmacro maj_L (nl) '(incf L (+ 3 ,nl)))
```

The initial unbound state is achieved by making each variable point to itself. The maj_L primitive updates the top of the *local stack* according to the size of the environment.

12.2.2.2. The Choice Case

Each choice block (Figure 12.3) owns a rerun part composed of the seven fields A, BCP, BCL, BG, BL, BP, TR (see section 7.2.5).

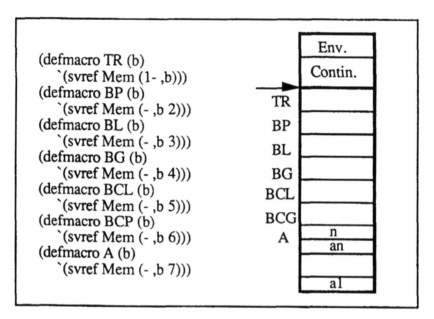

FIG. 12.3. Choice blocks

The A field purpose is to locate during backtracking the instance of the goal to be proved (see section 7.2.5). It is composed of the n+1 locations that describe the arity of the goal, followed by the n arguments of the goal.

During the creation of a choice point, the n arguments of the current goal should be saved, based on the Ai registers. All the arguments are pushed on top of the local zone, followed by the arity:

```
(defun save_args ()
  (dotimes (i (svref Mem A) (vset Mem (incf L i) i))
    (vset Mem (+ L i) (svref Mem (+ A i 1)))))
```

Upon each creation of a choice point, the arguments of the current goal are saved, together with the relevant registers.

```
(defun push_choix ()
  (save_args)
  (vset Mem (incf L) CP)
  (vset Mem (incf L) CL)
  (vset Mem (incf L) G)
  (vset Mem (incf L) BL)
  (vset Mem (incf L 2) TR)
  (setq BL (incf L) BG G))
```

Under this configuration, the rerun part can by itself provide all the necessary information for the backtracking (see section 7.2.5). The continuation and

environment parts could then be processed as in the deterministic case. We are
then left with the definition of the modification and deletion of a choice point:

```
(defun push_bpr (reste)
  (vset Mem (- BL 2) reste))
(defun pop_choix ()
  (setq L (- BL (size_C BL))
        BL (BL BL)
        BG (if (zerop BL) BottomG (BG BL))))
(defmacro size_C (b) '(+ 7 (A ,b)))
```

Unlike Mini-CProlog, the value of BG is directly held by the rerun part whose
size is no longer constant, but instead depends on the number of arguments to
be saved.

12.2.3. The *Trail*

The *trail* management is identical to that of Mini-CProlog (see section 11.2.4),
except for the restoring of variables:

```
(defmacro pushtrail (x)
  '(cond ((>= TR A) (throw 'debord (print "Trail Overflow")))
         ((vset Mem TR ,x) (incf TR))))

(defmacro poptrail (top)
  '(do () ((= TR ,top))
      (let ((v (svref Mem (decf TR))) )
        (vset Mem v (cons 'V v)))))
```

Indeed, the unbound state of a variable is now symbolized by the fact that it
"designates" itself.

12.3. Unification

The same presentation structure as for Mini-CProlog will be retained (see section 11.3): the access primitives will be described first, the unification algorithm
will be implemented next, and the binding mechanism will be considered last.

Unification in Mini-WAM differs from that in Mini-CProlog in two essential
points: first, it is associated with a terms representation by structure copying;
second, unification of a goal with a clause head is no longer performed in a
symmetrical fashion. Indeed, at each call of a goal, its arguments are saved
in the corresponding *Ai* registers (see section 7.2.3). The implications of such
choices will be described along with the presentation.

12.3.1. The Primitives

To access a structured term x requires taking into consideration two situations: either x is a model, in which case it should be considered within a given environment e (the `ultimate` function), or x is already a term instance (the `val` function) and the problem is readily solved.

```
(defun ultimate (x e)
  (if (var? x) (ult (adr x e)) x))
(defun val (x)
  (if (var? x) (ult (cdr x)) x))
(defmacro var? (x)
  '(and (consp ,x) (numberp (cdr ,x))))
(defmacro adr (v e) '(+ (cdr ,v) ,e))
```

In the case of a variable, the potential binding chains should be dereferenced. If the variable appears in a term model, we start from the location allocated to it in the environment (the `ultimate` function). If, instead, it appears in a term instance, then it provides itself the starting point (the `val` function). Dereferencing is performed by the `ult` function:

```
(defun ult (n)
  (let ((te (svref Mem n)))
    (if (and (var? te) (/= (cdr te) n)) (ult (cdr te)) te)))
```

12.3.2. The Unification Algorithm

Unification between a goal and the head of a clause is directly performed based on the *Ai* registers' holding the arguments of the goal. The `unify` definition (see section 11.3.2) should then be replaced by that of `unify_with`:

```
(defun unify_with (largs e)
  (catch 'impossible
    (do ((i (1+ A) (1+ i)))
        ((null largs))
      (unif (svref Mem i) (ultimate (pop largs) e)))))
```

Each *i*th argument of the head of the clause (i.e., *i*th car of `largs` to consider in the duplication environment e) is unified to the corresponding *Ai* register. Unlike in Mini-CProlog, unification is no longer a symmetrical operation (see section 11.3.2).

The `unif` function performs the unification between two terms x and y. For that matter, the equality between two constants is first considered. In the case of two different variables, they should be bound while still preserving the binding direction. In the case of a variable and a term, a classical binding is performed. Two different constants yield a failure. Finally, in the case of two structured terms having a single functional symbol and an equal arity, their arguments are unified:

```
(defun unif (x y)
  (cond
    ((eql x y) t)
    ((var? y) (if (var? x)
                  (if (= (cdr x) (cdr y)) t (bindv y x))
                  (bindte (cdr y) x)))
    ((var? x) (bindte (cdr x) y))
    ((or (atom x) (atom y)) (throw 'impossible 'fail))
    ((let ((b (copy? y)) (dx (pop x)) (dy (pop y)))
       (if (and (eq (functor dx) (functor dy))
                (= (arity dx) (arity dy)))
           (do () ((null x))
             (unif (val (pop x))
                   (if b
                       (val (pop y))
                       (ultimate (pop y) (E L)))))
           (throw 'impossible 'fail))))))
```

The non-symmetric aspect of unification appears again in the behavior of unif with respect to structured terms: since its first parameter x always appears in a goal argument (call of unif within unify_with), x is undoubtedly a term instance (access through the val function). However, since y appears in the head of a clause, it should be determined whether it is a model (access through ultimate) or a term instance made accessible through a variable (processing through val).

12.3.3. The Binding Mechanism

The bindv primitive performs the binding between two unbound variables while still maintaining their creation times. The bindte function processes the binding between a variable in the xadr location and a term y:

```
(defmacro bindv (x y)
  '(if (< (cdr ,x) (cdr ,y))
       (bind (cdr ,y) ,x)
       (bind (cdr ,x) ,y)))
(defun bindte (xadr y)
  (if (or (atom y) (copy? y))
      (bind xadr y)
      (bind xadr (recopy y (E L)))))
```

In the case of a model of a structured term, the binding goes through a copying using the current environment (see section 12.1.2). The bind function then performs the actual binding while memorizing it if it is postbound (see section 4.3.2.):

```
(defmacro bind (x te)
  '(progn
     (if (or (and (> ,x BottomL) (< ,x BL)) (<= ,x BG))
```

```
      (pushtrail ,x))
  (vset Mem ,x ,te)))
```

A variable is prior to the last choice point if it is local and prior to BL, or
in the copy stack and below BG (see section 6.3.5). Unlike Mini-CProlog, the
testing of BG (see section 11.3.3) is a "less than or equal to" one: the *global
stack* management relies on a top of stack register that indicates the last used
memory location instead of the first free one.

12.4. Control

The implementation of Mini-WAM will follow the same approach as that of
Mini-CProlog (see section 11.4): forward process and backtracking will be
described successively. Special attention will be paid to last-call optimization
during the implementation of the forward process. The goal arguments saving
mechanism will then be detailed.

12.4.1. The Forward Function

The forward process selects the current goal and executes it:

```
(defun forward ()
  (do () ((null Duboulot) (format t "no More ~%"))
    (cond ((null CP) (answer))
          ( (load_PC)
           (cond
             ((user? PC)
              (let ((d (def_of PC)))
                (if d (pr2 d) (backtrack))))
             ((builtin? PC)
              (if (eq (apply (car PC) (cdr PC)) 'fail)
                  (backtrack)
                  (cont_eval)))
             ((backtrack)))))))
(defun load_PC ()
  (setq PC (pop CP) PCE (E CL) Cut_pt BL))
```

Even though this definition implies a behavior identical to that of the forward
function in Mini-CProlog (see section 11.4.1), there exist two differences: pr2
replaces pr in order to apply the last-call optimization, and cont_eval is called
instead of cont_det after each successful execution of a built-in predicate.
More on this will follow in chapter 14.

The load_PC primitive loads the PC and PCE registers based on the current
continuation (CP,CL). Unlike Mini-CProlog (see section 11.4.1), this loading is
performed by modifying CP.

12.4.2. Last-Call Optimization

The last-call optimization will be implemented after describing the saving of the arguments of a goal. The application conditions will thus be verified accordingly and the *local stack* updating will be described.

12.4.2.1. Saving the Arguments of a Goal

Each argument of the current goal PC to be considered in the current environment PCE is saved in the corresponding *Ai* register. Four cases are to be considered: for a constant or an instance of a structured term, Ai is loaded with the corresponding value. For a term model, a copy on the top of the *global stack* is first performed (see section 7.2.3), then the register is loaded with this value. Finally, in the case of an unbound variable, the variable is dynamically reallocated in the global part if it is unsafe (see section 7.2.3).

```
(defun load_A (largs el)
  (dotimes (i (length largs) (vset Mem A i))
    (vset Mem
          (+ A i 1)
          (let ((te (ultimate (pop largs) el)))
            (cond
              ((atom te) te)
              ((var? te)
               (if (unsafe? te) (genvar (cdr te)) te))
              ((copy? te) te)
              ((recopy te el)))))))
(defun unsafe? (x)
  (and (not CP) (>= (cdr x) CL)))
```

The unsafe variables are those that occur in an unbound state in a goal in last position, and whose associated memory location appears in the local part of the environment that is to disappear, namely (E CL) in this case.

12.4.2.2. Implementation

The last-call optimization is performed by the pr2 function before effectively starting the current goal proof. It saves the arguments of the goal in the *Ai* registers, verifies the two application conditions (see section 7.1.1), and updates the *local stack* if the conditions are verified:

```
(defun pr2 (paq)
  (load_A (largs PC) PCE)
  (if CP
      (pr paq)
      (progn
        (if (<= BL CL) (setq L CL))
        (setq CP (CP CL) CL (CL CL))
        (pr paq))))
```

The optimization applies if the call is in last position (CP is empty) and no intermediate choice point has been created. The *local stack* is then updated (L:=CL) thus clearing the initial calling block. In order to determine the presence of intermediate choice points, the relative positions of the last choice block (BL) and those of the calling block (CL) are compared.

Note that this test also embeds the equality case. Indeed, turning each rerun part completely autonomous (see section 7.2.5) permits the discarding of the continuation and environment parts in the case where the caller is itself a choice point. Since the address of a choice point is located at the beginning of its continuation part (the rerun part is accessed through a negative offset; refer to section 12.2.2), the optimization also applies in the equality case. Accordingly, the continuation and environment parts are discarded while the rerun part remains.

The CP and CL registers should then be updated (in the case of a last-position call) so that they always designate the next sequence of goals to be proved and its associated block (see section 7.2.2).

12.4.3. The Proof

As for the previous implementation (see section 11.4.1), pr gives rise to a deterministic proof, or a choice proof, depending on the size of the clauses set. In the second case, the rerun part of a choice block is created on the top of the *local stack*.

```
(defun pr (paq)
  (if (cdr paq)
      (progn (push_choix) (pr_choice paq))
      (pr_det (car paq))))
```

12.4.3.1. Deterministic Case

The unification between the current goal (Ai registers) and the head of the clause is first achieved in the duplication environment:

```
(defun pr_det (c)
  (if (eq(unify_with(largs(head c))(push_E(nvar c))) 'fail)
      (backtrack)
      (when (tail c)
        (push_cont) (setq CP (tail c) CL L) (maj_L (nvar c)))))
```

In the case of a success, two situations can appear on the *local stack*:

1. The clause is a rule, thus the local block should be terminated: the current continuation is first saved before being updated (the body of the rule and the newly created block);

2. The clause is a fact: the local block is not terminated and L retains its initial value; unlike the precedent version (see section 11.4.1), it is not necessary to trace the different blocks in order to locate the next sequence of goals to be proven. This sequence is readily available from the CP and CL registers.

12.4.3.2. The Choice Case

As for Mini-CProlog (see section 11.4.1), the first clause whose head satisfies unification should be fetched first:

```
(defun shallow_backtrack (paq)
  (if (and (cdr paq)
           (eq (unify_with
                  (largs (head (car paq)))
                  (push_E (nvar (car paq))))
               'fail))
      (progn
        (poptrail (TR BL))
        (setq G BG)
        (shallow_backtrack (cdr paq)))
      paq))
```

Unlike Mini-CProlog, the unification is now performed through unify_with (see section 12.3.2). Moreover, the copying stack should be updated for each failure: its allocation is performed dynamically as the unification algorithm evolves, whereas the *global stack* was only updated after a success in the structure sharing solution (see section 11.4.1).

The pr_choice function is associated with the nondeterministic case. For that, it analyzes the result of the shallow_backtrack function and shifts to the deterministic case if there is only one clause left to be considered (see section 11.4.1).

```
(defun pr_choice (paq)
  (let* ((resu (shallow_backtrack paq))
         (c (car resu))
         (r (cdr resu)))
    (cond ((null r) (pop_choix) (pr_det c))
          ( (push_bpr r)
            (when (tail c)
              (push_cont)
              (setq CP (tail c) CL L)
              (maj_L (nvar c)))))))
```

In case of a nondeterministic success, the BP field of the rerun part should be updated. Then, two situations are possible for the *local stack*, depending on the type of clause:

1. c is a rule: the local block should be terminated. The old continuation is then saved accordingly before considering the new one.

2. c is a fact: the local block is not terminated and L remains unchanged. Indeed, we can be content with the rerun part since it now contains all the necessary information for restarting the forward process (see section 7.2.5). Here again it is not necessary to search for the next sequence of goals to be proved since it is readily available from the CP and CL registers.

12.4.4. Backtracking

The proof ends in the absence of a choice point. Otherwise, the computational status should be restored before restarting the forward process again:

```
(defun backtrack ()
  (if (zerop BL)
      (setq Duboulot nil)
      (progn
        (setq L BL G BG Cut_pt (BL BL)
              CP (BCP L) CL (BCL L))
        (load_A2)
        (poptrail (TR BL))
        (pr_choice (BP L)))))

(defun load_A2 ()
  (let ((deb (- L (size_C L))))
    (dotimes (i (A L) (vset Mem A i))
      (vset Mem (+ A i 1) (svref Mem (+ deb i))))))
```

Accordingly, each register recovers its initial value. Unlike Mini-CProlog (see section 11.4.2), all the information is gathered from the rerun part. In particular, continuation is obtained through the BCP and BCL fields, and the instance of the goal is obtained by restoring the A_i registers. For that, the n arguments are successively transferred before setting the arity.

What is left then is only to restore the environments and to restart the proof with the remaining clauses set.

12.5. Extensions

Let us conclude by suggesting some ideas for improvements.

1. Implement the trimming (see section 9.1.3).

   The parser should first be modified so that it maps an adequate numbering scheme to the variables of a clause C. Moreover, the coding of each literal on the body of a rule should explicitly embed the size of the environment to be retained during its call. Finally, the control should be modified in order to verify the trimming application conditions, and if needed, to perform it by updating the top of the *local stack* L. Special attention should be paid to the concept of unsafe variables that would thus appear (see section 9.1.3).

2. As for Mini-CProlog, make the local and copy stacks converge, in order to better allocate resources independently from the category of the programs.

CHAPTER 13

MINI·PROLOG·II

Mini-Prolog-II is a Prolog interpreter that corresponds to the third level, as described in Part 2. Accordingly, it implements the proposals of chapter 10. The Mini-Prolog-II characteristics are as follows: a terms representation by structure sharing, a four-stacks working zone architecture, an updating of the *local stack* by last-call optimization, and the implementation of the *dif*/2 and *freeze*/2 predicates.

Consequently, Mini-Prolog-II can be considered as equivalent to the interpreted version of Warren's Abstract Machine, except for the use of structure sharing (see chapter 7), and the addition of the *dif*/2 and *freeze*/2 predicates of Prolog-II (see chapter 10). Of course, Mini-Prolog-II will benefit from the clauses indexing mechanism presented in chapter 8.

This third implementation shares some characteristics with the two previous ones: the terms representation by structure sharing of Mini-CProlog (see chapter 11), and the blocks architecture, control, and last-call optimization of Mini-WAM (see chapter 12). The new contribution of Mini-Prolog-II is in the implementation of the *dif*/2 and *freeze*/2 predicates (see chapter 10).

This presentation will dissociate the aspects relative to the last-call optimization (see chapter 7 for the principle and chapter 12 for the implementation) from those specific to the *dif*/2 and *freeze*/2 predicates (see chapter 10). Accordingly, the unification algorithm will be implemented while taking into account the binding mechanism of frozen variables. The control will be implemented next by following the same procedure as for Mini-WAM (see section 12.4), and the specific aspects of the implementation of *dif*/2 and *freeze*/2 will be described.

Finally, note that the Mini-Prolog-II code is divided into two parts: a common part it shares with Mini-CProlog (listed in appendix D) and its intrinsic part (listed in appendix C).

13.1. The Working Zone Architecture

The global architecture of the working zone will first be described while focusing on the nature of the Ai registers in the context of structure sharing. Next, the form of the activation blocks and the updating primitives of each of the four stacks will be described successively.

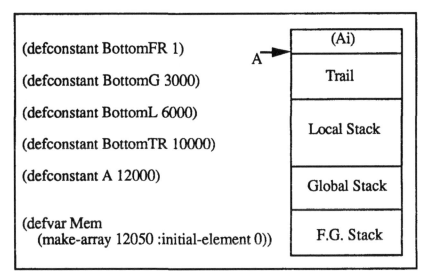

FIG. 13.1. Working area architecture

13.1.1. General Architecture

As for Mini-WAM (see section 12.2.1), the Ai registers are represented as a vector located on the top of the working zone. Unlike Mini-WAM, each Ai register will be a pair (see section 7.2.3), since structure sharing is used.

The working zone is then divided into five parts (see section 10.3.1), organized as depicted in Figure 13.1.

The working registers are identical to those of Mini-WAM (see section 12.2.1), and specific registers for the management of frozen goals have been added:

- TR: top of the *trail*
- L: top of the *local stack*
- G: top of the *global stack*
- FR: top of the stack of frozen goals
- CP: Continuation point
- CL: local block of CP
- BL: local block of the last choice point
- BG: top of the *global stack* associated with the last choice point.
- PC, PCE, PCG: the current goal in both its environments.
- FRCP: list of awakened goals.

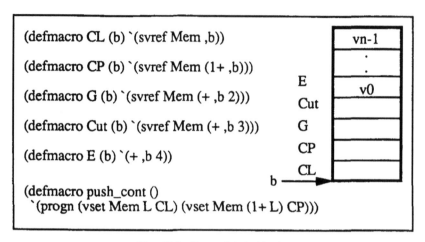

FIG. 13.2. Deterministic blocks

As for Mini-CProlog, each top of stack register will designate the first free location.

13.1.2. The *Local Stack*

The structure of the activation blocks is identical to that of Mini-WAM (see section 12.2.2). Indeed, it results from the same objective: implementing the last-call optimization (see section 7.2.3).

However, there exist two small differences with respect to Mini-WAM. First, clause duplication results in the division of the allocated environment according to its local and global parts. Next, each rerun part has an additional field FR (see section 10.3.5) that describes the top of the stack of frozen goals associated with the last choice point: this is inherent in the implementation of the delaying mechanism.

13.1.2.1. The Deterministic Blocks

Each deterministic block is decomposed according to its continuation and environment parts (see Figure 13.2).

During the duplication of a clause that holds n local variables, the push_E primitive allocates an environment with a corresponding size on the top of the *local stack*.

```
(defmacro push_E (n)
  `(let ((top (+ L 4 ,n)))
     (if (>= top BottomTR)
         (throw 'debord (print "Local Stack Overflow")))
     (vset Mem (+ L 3) Cut_pt)
```

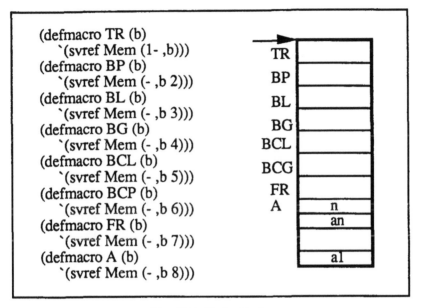

```
(defmacro TR (b)
  `(svref Mem (1- ,b)))
(defmacro BP (b)
  `(svref Mem (- ,b 2)))
(defmacro BL (b)
  `(svref Mem (- ,b 3)))
(defmacro BG (b)
  `(svref Mem (- ,b 4)))
(defmacro BCL (b)
  `(svref Mem (- ,b 5)))
(defmacro BCP (b)
  `(svref Mem (- ,b 6)))
(defmacro FR (b)
  `(svref Mem (- ,b 7)))
(defmacro A (b)
  `(svref Mem (- ,b 8)))
```

FIG. 13.3. Choice blocks

```
(dotimes (i ,n top)
    (vset Mem (decf top) (cons 'LIBRE BottomG)))))
(defmacro maj_L (nl) '(incf L (+ 4 ,nl)))
```

Unlike Mini-CProlog (see section 11.2.2), and as for Mini-WAM (see section 12.2.2), the last choice point prior to the call (the *cut* field) is saved. Finally, maj_L updates the top of the *local stack* according to the size of the allocated environment.

13.1.2.2. The Choice Blocks

The rerun part (see Figure 13.3) of a choice block is made up of eight fields, seven of which are also found in Mini-WAM: A, BCP, BCL, BG, BL, BP, TR (see section 12.2.2).

The additional field FR (see section 10.3.5) describes the top of the stack of frozen goals at the creation of the choice point. It will be used for the updating of this stack, upon backtracking.

As for Mini-WAM, each time a choice point is created, the working registers should be saved and the n arguments of the calling goal should be pushed in order to be able subsequently to restart the forward process. The updating of the BP field and the deletion of a choice point are performed as in the Mini-WAM (see section 12.2.2).

```
(defun save_args ()
  (dotimes (i (svref Mem A) (vset Mem (incf L i) i))
    (vset Mem (+ L i) (svref Mem (+ A i 1)))))
(defun push_choix ()
  (save_args)
  (vset Mem (incf L) FR)
  (vset Mem (incf L) CP)
  (vset Mem (incf L) CL)
  (vset Mem (incf L) G)
  (vset Mem (incf L) BL)
  (vset Mem (incf L 2) TR)
  (setq BL (incf L) BG G))
```

13.1.3. The *Global Stack*

Since the implementation of the last-call optimization does not affect the behavior of the *global stack*, its management will be the same as for Mini-CProlog (see section 11.2.3).

Note that the initial concept of a global variable should be extended to potentially unsafe ones (see section 7.2.3). Finally, any variable that appears in a *dif* or *freeze* will become global in order to preserve the properties of the *local stack* (see section 10.3.5). The parser will be modified accordingly (see appendix C).

13.1.4. The *Trail*

The trail management should be modified to take into account the frozen variables (see section 10.3.5). Restoring a variable is then performed accordingly: if it is a frozen variable, it is restored to the pair (LIBRE, frozen block).

```
(defun pushtrail (x)
  (if (>= TR A) (throw 'debord (print "Trail Overflow")))
  (vset Mem TR x)
  (incf TR))
(defun poptrail (top)
  (do () ((= TR top))
    (let ((x (svref Mem (decf TR))))
      (if (numberp x)
          (vset Mem x (cons 'LIBRE BottomG))
          (vset Mem (car x) (cons 'LIBRE (cdr x)))))))
```

13.1.5. The Stack of Frozen Goals

Each frozen block is made up of its four fields, FGvar, FGtail, FGgoal, and FGenv (see section 10.3.1):

```
(defmacro FGvar (x) '(svref Mem ,x))
(defmacro FGtail (x) '(svref Mem (1+ ,x)))
(defmacro FGgoal (x) '(svref Mem (+ 2 ,x)))
```

```
(defmacro FGenv (x) '(svref Mem (+ 3 ,x)))

(defmacro fgblock (x) '(cdr (svref Mem ,x)))
(defmacro frozen? (x) '(< (fgblock ,x) BottomG))
```

A frozen unbound variable is characterized by the fact that, in the pair that
represents it, the second element designates a block on the stack of frozen goals
(see section 10.3.2). Since the latter is located directly beneath the *global stack*,
the comparison is made with the value BottomG.

```
(defmacro push_fg (v b eb r)
  '(if (>= (+ FR 3) BottomG)
       (throw 'debord (print "Frozen Goals Stack Overflow"))
       (progn (vset Mem FR ,v)
              (vset Mem (incf FR) ,r)
              (vset Mem (incf FR) ,b)
              (vset Mem (incf FR) ,eb)
              (incf FR))))
```

The allocation of a frozen block is performed through the primitive push_fg,
which successively pushes each of the four values to be saved: the variable v,
the (potentially empty) sequence re of already delayed goals, and the new goal
b with its global environment eb.

13.2. Unification

Since Mini-Prolog-II uses structure sharing, the implementation of its unification algorithm reuses many primitives already defined for Mini-CProlog (see
section 11.3). We will describe here the two enhancements needed by the last-
call optimization (unification with the Ai registers) and by the implementation
of the delaying mechanism.

13.2.1. The Unification Algorithm

Unlike Mini-CProlog, the arguments of a goal are loaded in the Ai registers.
Consequently, the unification between a goal and the head of a clause will now
be performed by the unify_with function:

```
(defun unify_with (largs el eg)
  (catch 'impossible
    (dotimes (i (svref Mem A))
      (unif
        (let ((te (svref Mem (+ A 1 i))))
             (val (car te) (cdr te)))
        (ultimate (pop largs) el eg)))))
```

Each ith argument of the head of the clause (ith car of largs to consider in
the (el, eg) duplication environments) is unified with the corresponding Ai

register. The other primitives (`unif`, `ultimate`, `val`, `ult`, `value`, `adr`) remain unchanged with respect to Mini-CProlog (see section 11.3).

13.2.2. Binding Algorithm

Each time a frozen variable is bound (to a constant, another variable, or a structured term), its sequence of delayed goals should be located in order to be thawed after a successful unification (see section 10.3). Accordingly, the `bindte` function differentiates between two cases.

```
(defun bindte (x sq e)
  (if (frozen? x)
      (let ((y (fgblock x)))
         (push y FRCP)
         (bind x sq e (cons x y)))
      (bind x sq e x)))
(defmacro bind (x sq e xt)
  '(progn
     (if (or (and (> ,x BottomL) (< ,x BL)) (< ,x BG))
         (pushtrail ,xt))
     (rplaca (svref Mem ,x) ,sq)
     (rplacd (svref Mem ,x) ,e)))
```

In the case of a frozen unbound variable, its frozen block (which provides access to the sequence of delayed goals) is added in front of the current list FRCP of thawed goals. Additionally, the *trail* should memorize the variable together with its frozen block for potential backtracks (see section 10.3.5). In the case of a truly unbound variable, the binding is performed normally (see section 11.3.3).

The `bindv` primitive, which determines the binding direction between two unbound variables, remains unchanged (see section 11.3.3). After comparing their creation times, the binding is performed by calling `bindte`.

```
(defmacro bindv (x ex y ey)
  '(let ((ax (adr ,x ,ex)) (ay (adr ,y ,ey)))
     (if (< ax ay)
         (bindte ay ,x ,ex)
         (bindte ax ,y ,ey))))
```

13.3. Control

The outline for Mini-WAM will be followed here again (see section 12.4). Indeed, the structure of the activation blocks and the control management is identical (apart from the terms representation) to those of the second implementation. The modifications necessary for the implementation of the delaying mechanism (see section 10.3.5) will be described throughout this exposé.

13.3.1. The Definition of the Forward Function

The forward process selects the current goal and executes it:

```
(defun forward ()
  (do () ((null Duboulot) (format t "no More~%"))
    (cond
      ((and (null CP) (null FRCP)) (answer))
      ((load_PC)
       (cond
         ((user? PC)
          (let ((d (def_of PC)))
            (if d (pr2 d) (backtrack))))
         ((builtin? PC)
          (if (eq (apply (car PC) (cdr PC)) 'fail)
              (backtrack)
              (cont_eval)))
         ((backtrack)))))))
```

Unlike Mini-WAM (see section 12.4.1), the potential sequences of goals thawed by the last unification should be taken into account. These sequences are held by the FRCP register (see section 13.1.1 and 13.1.2). Consequently, a proof ends when CP and FRCP are both empty. The load_PC primitive still performs the loading of the PC, PCE, and PCG registers. However, it also manages the awakened goals (see section 13.4.4).

13.3.2. The Last-Call Optimization

The load_A primitive performs the loading of the arguments of the goal PC to be considered in the environments (PCE, PCG). Unlike Mini-WAM (see section 12.4.2), no problem can occur due to unsafe variables. Indeed, this difficulty has been statically resolved by considering as global any potentially unsafe variable (see section 13.1.3).

```
(defun load_A (largs el eg)
  (dotimes (i (length largs) (vset Mem A 1))
    (vset Mem (+ A i 1) (ultimate (pop largs) el eg))))
```

The last-call optimization is then performed as for Mini-WAM (see section 12.4.2):

```
(defun pr2 (paq)
  (load_A (largs PC) PCE PCG)
  (if CP
      (pr paq)
      (progn
        (if (<= BL CL) (setq L CL))
        (setq CP (CP CL) CL (CL CL))
        (pr paq))))
```

For a call in the last position, and in the absence of intermediate choice points, the *local stack* is updated, thus leading to the discarding of the father block.

13.3.3. The Proof

In the deterministic case, unification is first performed between the current goal (the Ai registers) and the head of the clause in its duplication environment:

```
(defun pr_det (c)
  (if (eq (unify_with
            (largs (head c))
            (push_E (nloc c))
            (push_G (nglob c)))
         'fail)
      (backtrack)
      (progn
        (maj_G (nglob c))
        (when (tail c)
          (push_cont)
          (setq CP (tail c) CL L)
          (maj_L (nloc c)))))))
```

Unlike Mini-WAM, the duplication environment is divided into local and global parts (see section 12.4.3). Accordingly, the *global stack* should be updated in case of success.

In the case of a nondeterministic proof, the first clause whose head satisfies unification should be picked first. This is the purpose of shallow_backtrack:

```
(defun shallow_backtrack (paq)
  (if (and (cdr paq)
           (eq (unify_with
                 (largs (head (car paq)))
                 (push_E (nloc (car paq)))
                 (push_G (nglob (car paq))))
              'fail))
      (progn
        (setq FRCP nil FR (FR BL))
        (poptrail (TR BL))
        (shallow_backtrack (cdr paq)))
      paq))
```

In the case of a failure, the prior status of bindings is restored. The stack of frozen goals should also recover its initial status. Finally, the different goals thawed by the unification that just produced a failure should not be inadvertently awakened.

As for Mini-WAM (see section 12.4.3), the pr_choice function analyzes the result of shallow_backtrack and turns to the deterministic case if there is only one clause left to consider:

```
(defun pr_choice (paq)
  (let* ((resu (shallow_backtrack paq))
         (c (car resu))
         (r (cdr resu)))
    (cond
     ((null r) (pop_choix) (pr_det c))
     ( (push_bpr r)
       (maj_G (nglob c))
       (when (tail c)
             (push_cont)
             (setq CP (tail c) CL L)
             (maj_L (nloc c)))))))
```

In case of success, the *global stack* is updated. The forward process then continues and differentiates between two cases (see section 12.4.3) depending on whether the clause c is a fact or a rule.

13.3.4. Backtracking

The proof ends in the absence of choice points. Otherwise, the computational status should be restored before restarting the forward process:

```
(defun backtrack ()
  (if (zerop BL)
      (setq Duboulot nil)
      (progn
        (setq L BL G BG FR (FR L) FRCP nil Cut_pt (BL BL)
              CP (BCP L) CL (BCL L) Cut_pt (BL BL))
        (load_A2)
        (poptrail (TR BL))
        (pr_choice (BP L)))))
```

The registers relevant to classical control are thus updated, including the two registers specific to the delaying mechanism: FRCP and FR.

13.4. The Delaying Mechanism

The explanatory sequence will be the same as that of section 10.3. The *dif*/2 and *freeze*/2 predicates will be implemented after describing the association between a variable and its sequence of delayed goals. Next, we will focus on the awakening strategy, in which the *dif*(*s*) will always be proved first, and the other goals will be proved according to a queue scheduling à la MU-Prolog. The implementation of the built-in predicate frozen_goals/0 (which determines the set of unthawed delayed goals and their associated variables) will come next.

13.4.1. The Binding of Variables and Delayed Goals Sequences

The binding between a variable and a sequence of delayed goals is performed by the `bindfg` primitive (see section 10.3.2):

```
(defun bindfg (x b eb r)
  (bind x 'LIBRE FR (if (frozen? x) (cons x r) x))
  (push_fg x.b eb r))
```

A new block is allocated on the top of the stack of frozen goals in order to memorize the new delaying. The association with the variable is performed by making the second element of its pair point to the block just created. As for any true binding of variables, this association should be recorded if it is postbound.

13.4.2. The *freeze*/2 Predicate

The execution of a goal `freeze(X,P)` depends on the status of the binding of the X variable (see section 10.3.2). If X is bound to a constant or a structured term, then P is executed. If X is an unbound variable, then P is delayed. The freezing is performed through the call of the `bindfg` function with, as arguments: y (the dereferenced value of x), the potential frozen block already associated with y, and the goal itself together with its execution environment.

```
(defun |freeze| (x p)
  (let ((xte (ultimate x PCE PCG)))
    (if (var? (car xte))
        (let ((y (adr (car xte) (cdr xte)))
              (pte (ultimate p PCE PCG)))
          (bindfg y (dec_goal (car pte))
                    (cdr pte) (fgblock y)))
        (|call| p))))
```

The cases where y is a truly unbound variable or a frozen unbound one (see section 10.3.2) are merged into a single one, taking into account the environments initialization (see section 13.1.2 and 13.1.3). Indeed, each pair representing an unbound variable has the form (LIBRE, BottomG), so any unbound variable is automatically associated with an empty sequence of delayed goals (see the definition of `frozen?` in section 13.1.5).

13.4.3. The *dif*/2 Predicate

Since the effects of the unification attempted by *dif*/2 should not prevail, each variable binding must be memorized on the *trail* (see section 10.3.2). Accordingly, the BL and BG registers will temporarily designate the current tops of the *local* and *global stacks*. The current top of the *trail* is also saved:

```
(defun |dif| (x y)
  (let ((BL L) (BG G) (str TR) (FRCP nil))
    (if (eq (uni x y) 'fail)
        (poptrail str)
        (if (/= TR str)
            (let* ((xv (svref Mem (1- TR)))
                   (v (if (numberp xv) xv (car xv))))
              (poptrail str)
              (bindfg v PC PCG (fgblock v)))
          'fail))))
```

Three cases are then to be distinguished, depending on the result of unification between x and y:

1. failure: dif(x,y) succeeds, and the bindings status should be restored;

2. success without a variable binding: dif(x,y) fails;

3. success with the binding of at least one variable v: dif(x,y) is delayed on v. The status of the bindings is restored, and the variable v is associated with the new sequence of delayed goals.

Determining the variable v, by inspecting the last block of the *trail*, should take into account the case of a frozen unbound variable (see section 13.2.2), that is, the case where the pair (variable, delayed block) was memorized.

13.4.4. Triggering Strategy

The presence of thawed goals should be determined before each execution of a new goal. If a thawed goal is found, its proof should be performed immediately. This process is performed by the load_PC primitive, each time the forward function is called (see section 13.3.1):

```
(defun load_PC ()
  (if FRCP
      (let ((x ()))
        (do () ((null FRCP)) (setq x (add_fg (pop FRCP) x)))
        (do () ((null x)) (create_block (abs (pop x)))))))
  (setq PC (pop CP) PCE (E CL) PCG (G CL) Cut_pt BL))
```

In a case where thawed goals are found (FRCP is not empty), they are first assembled into a list where ordering depends on their freezing times (see section 10.2.5). Next, they are pushed on the top of the *local stack* to be immediately executed.

We have decided to implement a sort by decreasing order, since the pushing will be performed by a sequential left-to-right scanning of the list just constructed. Consequently, the goal with the oldest freezing time (the first to be delayed) will be triggered first (queuing strategy of section 10.3.3).

```
(defun insert (b 1)
  (if (or (null 1) (> b (car 1)))
      (cons b 1)
      (cons (car 1) (insert b (cdr 1))))))
```

Finally, note that in order to implement the IC-Prolog approach (see section 10.2.5), it is sufficient to perform a sort in increasing order since the pushing reverses this order. With respect to the proposed version, only the comparison test in the definition of insert needs to be inverted.

13.4.4.1. The Triggering Order

The implemented awakening strategy is that which is described in section 10.3.3: the *dif(s)* first, followed by the other goals depending on their freezing times. Delaying is perpetuated if the awakening is due to a binding between two unbound variables.

The add_fg function examines the sequence of goals associated with a variable (starting from its frozen block b) while inserting each new goal in the ordered list r of goals already encountered:

```
(defun add_fg (b r)
  (let ((b1 (if (numberp (FGgoal b)) (FGgoal b) b)))
    (if (eq (pred (FGgoal b1)) '|dif|)
        (insert (- b1) (other_fg b r))
        (let* ((v (svref Mem (FGvar b1)))
               (te (val (car v) (cdr v))))
          (if (var? (car te))
              (let ((y (adr (car te) (cdr te))))
                (bindfg y b1 nil (fgblock y))
                (other_fg b r))
              (insert b1 (other_fg b r)))))))
(defun other_fg (b r)
  (if (< (FGtail b) BottomG) (add_fg (FGtail b) r) r))
```

Three cases can appear for a given block b1:

1. It is a dif: a freezing time older than that of any other goal is associated with it;

2. The (FGvar b1) variable is unbound: delaying is perpetuated on the dereferenced value;

3. The (FGvar b1) variable is bound to a constant or a structured term: the goal is awakened according to its freezing time b1.

In order to preserve the true freezing times (see section 10.3.3) during another delaying, the FGgoal field will no longer directly designate the goal but rather the frozen block where the goal appears. Consequently, upon each call of (add_fg b r), (FGgoal b) should be dereferenced in order to access the initial frozen block.

13.4.4.2. The Blocks Allocation

For each block b holding a goal (FGgoal b) in its global execution environment (FGenv b), a new deterministic block is allocated on the *local stack*:

```
(defun create_block (b)
  (push_cont)
  (vset Mem (+ L 2) (FGenv b))
  (vset Mem (+ L 3) Cut_pt)
  (setq CP (list (FGgoal b)) CL L)
  (maj_L 0))
```

Accordingly, the current continuation is saved (push_cont). Next, an environment is created (artificially): the field G (which has the address L+2) designates the global execution environment of the goal, and the E field (with an offset of 4 with respect to L) is an empty local environment. The two current continuation registers CP and CL should then be updated so that they designate the goal to be proved in the block just created. What is left then is to update the top of the *local stack*.

13.4.5. The Built-in Predicate *frozen-goals/0*

The built-in predicate frozen_goal/0 determines any time during the proof the set of delayed and still-not-awakened goals, together with their associated variables (see section 10.3.4).

```
(defun |frozen_goals| ()
  (do ((i (- FR 4) (- i 4)))
      ((< i 0))
    (if (eq (car (svref Mem (FGvar i))) 'LIBRE)
        (let ((b (if (numberp (FGgoal i)) (FGgoal i) i)))
          (writesf (pred (FGgoal b))
                   (largs (FGgoal b))
                   (FGenv b))
          (format t " frozen upon X~A~%" (FGvar i))))))
```

The stack of frozen goals is sequentially scanned to detect whether for each block i its (FGvar i) is still in its initial unbound state. If it is, then the goal designated by the block i has not yet been awakened (see section 10.3.4).

In order to access the goal, its FGgoal field should be dereferenced (see section 13.4.4). The instance of the goal is then printed out, together with the corresponding frozen variable.

13.5. Extensions

Let us conclude with some ideas for improvements:

1. Complete Mini-Prolog-II to make it perform the unification on infinite terms. For that, the algorithm (see section 1.4.2) and the first implementation (see section 1.5.3) should be reconsidered;

2. Define the `bag_of_frozen_goals/1` predicate, which unifies its argument with the list of upon(Goal,Var) pairs, where Goal is a goal delayed on Var that has not yet been awakened.

    Such goals would then become directly processable by the user, who would be able to reactivate them himself through the bindings of the corresponding variables.

`frozen_goals/0` is expressed in terms of `bag_of_frozen_goals/1` in the following manner:

```
frozen_goals:-bag_of_frozen_goals(L),
    write_fg(L).
write_fg([]).
write_fg([upon(Goal,V)|Tail]):-write(Goal),
    write('frozen upon'),
    write(V),nl,
    write_fg(Tail).
```

CHAPTER 14

BUILT-IN PREDICATES

This chapter is relevant to the three versions just implemented. For the sake of homogeneity, an identical management of the built-in predicates has been chosen for the three implementations. Each version has its own built-in predicates (with a common part for Mini-CProlog and Mini-Prolog-II due to their common representation of terms). Moreover, each implementation has its own initialization file, which contains a set of utilities, themselves written in Prolog. The presentation will always refer to Mini-Prolog-II.

The built-in predicates can be classified into many different classes, depending on their purpose: I/O, control, arithmetic, type conversion, tests, term comparison, dynamic management of the clauses base, ... This presentation will only consider a subset of these. The list is far from being exhaustive, and the reader is highly encouraged to define new ones, based on the scheme of those presented in this chapter.

The management of the built-in predicates common to the three versions will be described first. Two I/O predicates, read/1 and write/1, will then be studied. The control part will be tackled next, and the implementation of the famous cut/0 predicate will be detailed. The usual arithmetic and type test predicates will then be presented. Finally, the dynamic knowledge base updating predicates will be described.

14.1. Management of the Built-in Predicates

The key idea of the implementation is to associate with each built-in predicate a Lisp function with a similar name and the same number of arguments. Each built-in predicate is marked and then recognized as such during the proof. Consider again the definition of the forward function of Mini-Prolog-II (see section 13.3.1):

```
(defun forward ()
  (do () ((null Duboulot) (format t "no More~%"))
      (cond ((and (null CP) (null FRCP)) (answer))
            ((load_PC)
             (cond
              ((user? PC)
               (let ((d (def_of PC)))
                 (if d (pr2 d) (backtrack))))
              ((builtin? PC)
```

```
(if (eq (apply (car PC) (cdr PC)) 'fail)
    (backtrack)
    (cont_eval)))
  ((backtrack)))))))
```

When a built-in predicate is encountered, its associated Lisp function is evaluated. The returned value is then analyzed in order to determine if the execution has lead to a success or a failure. The proof then evolves as a consequence.

The different built-in predicates are assembled into a list for their initial marking:

```
(defvar Ob_Micro_PrologII
  '( |write| |nl| |tab| |read| |get| |get0| |var| |nonvar|
     |atomic| |atom| |number| |fail| |true| ! |div1| |mod|
     |plus| |minus| |mult| |le| |lt| |name| |consult|
     |abolish| |cputime| |statistics| |call| |freeze| |dif|
     |frozen_goals|  ) )

(mapc #'(lambda (x) (setf (get x 'evaluable) t))
      Ob_Micro_PrologII)
```

What is left then is to solve the problem of a possible conflict with Lisp for the built-in predicates having the same name as a predefined function. Since Common-Lisp implicitly considers the atoms' names as uppercase, any function associated with a built-in predicate will have its name in lowercase.

```
(defmacro value1 (x)
  '(car (ultimate ,x PCE PCG)))
(defun uni (x y)
  (catch 'impossible
    (unif (ultimate x PCE PCG) (ultimate y PCE PCG))))
```

The value1 and uni primitives simplify the definitions of the functions associated with built-in predicates. First of all, value1 will allow the access to the dereferenced value of an argument. Finally, uni will perform the unification between two terms x and y through the call of the unif function (see section 11.3).

For each built-in predicate subsequently described, its arity will be indicated and the behavior of its arguments will be specified according to the mode declaration formalism of Prolog-Dec10 (see section 6.4.3). Thus, + will be associated with an input argument (instantiated during the call), − will be associated with an output argument (an unbound variable during the call), and ? will represent an argument that can behave both ways.

14.2. Input/Output

The implementation of the read/1 and write/1 predicates, whose respective functions are to read and to print Prolog terms, will be presented here. The Lisp definition of the other I/O predicates appears in appendix D.

14.2.1. The read/1 Predicate

The read/1 predicate reads the next Prolog term (always followed by the "end" marker ".") on the input stream and unifies it with its argument. Accordingly, the term is read, then coded:

```
; read/1 (?term)
(defun |read| (x)
  (let ((te (read_terme)))
    (catch 'impossible
      (unif (ultimate x PCE PCG)
            (cons (cdr te) (push1_g (car te)))))))
(defun read_terme ()
  (let ((*lvarloc nil) (*lvarglob nil))
    (let ((te (read_term (rchnsep) 2)))
      (rchnsep)
      (cons (length *lvarglob) (c te)))))
```

Any variable that appears in the te term thus read is considered as global. read/1 then unifies its argument to the te term, for which a new global environment is allocated.

14.2.2. The write/1 Predicate

The write/1 predicate prints its argument, independently of its type: empty list, constant, variable, nonempty list, or functional term:

```
; write/1 (?term)
(defun |write| (x)
  (write1 (ultimate x PCE PCG)))
(defun write1 (te)
  (let ((x (car te)) (e (cdr te)))
    (cond
      ((null x) (format t "[]"))
      ((atom x) (format t "~A" x))
      ((var? x) (format t "X~A" (adr x e)))
      ((list? x) (format t "[")
       (writesl (val (cadr x) e) (val (caddr x) e))
       (format t "]"))
      ((writesf (functor (des x)) (largs x) e)))))
```

In the case of an unbound variable, an external name based on its memory address is printed. This enables the identical representation of many occurrences of the same variable.

In the case of a nonempty list, each of its elements is written successively, while differentiating the case of a list and that of a dotted pair:

```
(defun writesl (te r)
  (write1 te)
  (let ((q (car r)) (e (cdr r)))
    (cond
      ((null q))
      ((var? q) (format t "|X~A" (adr q e)))
      (t (format t ",")
         (writesl (val (cadr q) e) (val (caddr q) e))))))
```

Finally, in the case of a functional term with an arity n, the functional symbol is printed, followed by its n arguments:

```
(defun writesf (fct largs e)
  (format t "~A(" fct)
  (write1 (val (car largs) e))
  (mapc #' (lambda (x) (format t ",") (write1 (val x e)))
        (cdr largs))
  (format t ")"))
```

14.3. Control

Two examples of control built-in predicates are presented here: the cut/0 and call/1 predicates.

14.3.1. The Predicate cut/0

The cut/0 predicate, denoted ! in the Edinburgh Prologs and / in the Marseille ones, allows for the control of the expansion of the search tree by eliminating choice points. It was first introduced by Alain Colmerauer in version 1 of Prolog developed at the University of Marseille [BM73], [Rou75].

After briefly recalling the definition of the *cut* predicate, we will describe its implementation and show how it fits the memory management of both the *local stack* and the *trail*. Finally, we will study the proposal of Prolog-Bordeaux [Bil85], [Bil86] which defines a new primitive of the if-then-else type, the cond/3 predicate, intended to replace the *cut*.

14.3.1.1. The Function of *cut*

The *cut* semantics is defined solely in terms of control and is quite tricky to formulate [Con86]: "when a cut is embodied in a rule, its execution forces the elimination of all the choice points for all the goals starting from the one that triggered the rule containing the *cut*, up to the one that precedes the cut in the body of this rule."

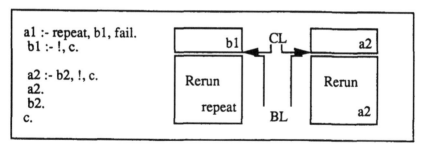

FIG. 14.1. The *cut* problem

An equivalent definition can be given in terms of a tree representation (see section 2.3): the *cut* eliminates all the *or* nodes (built up to that point) that appear in the subtree whose root is its father goal, including the latter if it is itself an *or* node. The effect of the *cut* is thus to turn the beginning of the current proof of its father goal deterministic.

14.3.1.2. Memory Management and Implementation

The implementation of *cut* consists in updating the backtracking registers BG and BL and exploiting the determinism that has thus appeared in order to recover memory on the *local stack* and the *trail*.

Following the application of the *cut*, the last active choice point should be the one preceding the call of *cut* father goal, whose block is designated by the register CL. Although in Mini-CProlog the correct value of BL is provided (by tracing the BL fields of the choice blocks) by the first value strictly lower than CL, it is not the same in Mini-WAM and Mini-Prolog-II.

Indeed, because of the autonomy of each rerun part during the implementation of the last-call optimization (see section 7.2.5), it becomes impossible to determine if the choice point designated by BL should or should not disappear in the equality case, that is, $BL = CL$.

$BL = CL$ appears during the execution of the cut triggered by b1/0 in the proof of a1/0, but the choice point on repeat/0 should not disappear (see Figure 14.1). However, in the proof of a2/0, the choice point should be *cut*, even though the two registers are equal.

This problem is solved in Mini-WAM (see section 12.2.2) and in Mini-Prolog-II (see section 13.1.2) by adding [BBCT86] a new field to each activation block.

This *cut* field memorizes the last choice point that precedes the call of the goal. Accordingly, a supplementary working register *cutpt* is updated upon each call of a new goal (see section 12.4.1 and section 13.4.4), and then restored to the correct value during backtracking.

```
;  !/0 (version Mini-CProlog)
(defun ! (n)
  (unless (< BL CL)
          (do () ((< BL CL)) (setq BL (BL BL)))
          (setq BG (if (zerop BL) BottomG (G BL))
                L (+ CL 3 n))))
;  !/0 (version Mini-PrologII)
(defun ! (n)
  (setq BL (Cut CL)
        BG (if (zerop BL) BottomG (BG BL))
        L (+ CL 4 n)))
```

Thus, for Mini-WAM and Mini-Prolog-II, the *cut* field of the father block directly provides the new value of BL. Finally, in all cases, the updating of the BG register is performed by accessing the new block just designated by BL.

However, since the execution of the *cut* turns the beginning of the current proof of its father goal deterministic, the *local stack* can consequently be updated. Indeed, it is sufficient to keep the father block on the top of the stack. Since CL indicates the address of the beginning of block, the updating of the L register should take into account the size of block, and thus the number of variables that appear in the local part of its environment (see section 13.1.2).

Accordingly, a parameter that indicates the size of the local part of the environment (i.e., the total number of variables for the structure copying, or the number of local variables for structure sharing) is associated with the *cut* during the static coding of the rule. What is left then is to update L while taking into account the number of fields that remain, specifically (CL, CP, G) for Mini-CProlog, (CL, CP, *cut*) for Mini-WAM, and (CL, CP, G, *cut*) for Mini-Prolog-II.

Finally, note that this introduction of determinism appears also at the level of the *trail* management. Indeed, the elimination of choice points renders useless the retention, in the *trail*, of all the variables with creation times higher than those of the (new) last choice point.

We propose, as an extension, to update the *trail* upon each call of the *cut* predicate. Accordingly, let $tr0$ be the top of the *trail* associated with the new last choice point: all the located variables (between $tr0$ and TR) on the *trail* should be considered, eliminating all those with a creation time higher than the last choice point: $(>= BL)$ for a local variable, or $(>= BG)$ for a global one. This updating is of course valid for the three versions.

14.3.1.3. Prolog-Bordeaux: A Prolog without *cut*

Prolog-Bordeaux [Bil85], [Bil86] proposes a new primitive of the algorithmic type analogous to the $if - then - else$ of classical languages: the cond/3 predicate that replaces the use of the *cut*. Considering the definition of the cond/3 predicate as it appears in [Bil85], the goal cond(Test,If_true,If_false) is evaluated as follows:

1. look for the first success of Test;

2. if such a success exists, then look for the successes of If_true compatible with Test;

3. if there is no success for Test, then look for the successes of If_false.

In the case where backtracking follows cond(Test,If_true,If_false), control returns to the eventual choice point associated with If_true (if Test succeeded), or associated with If_false (if Test failed). In no case will the first success of Test be reconsidered.

M. Billaud has shown [Bil85] that the power of Prolog without *or* but with *cut*, is similar to that of Prolog with *cond* but without *cut*. As a result, the *cut* can be removed from all Prolog programs (without *or*) and replaced with the *cond* primitive.

However, the power of Prolog with *cond*/3 remains weaker than Prolog with *or* [Bil86], because some programming schemes that combine *cut* and *or* cannot be translated in terms of *cond*.

14.3.2. The call/1 Predicate

In order to implement the call/1 predicate, two cases have to be distinguished, depending on whether its argument is a variable predicate. Indeed, if its argument is a variable, then the execution environment of the goal it represents is not necessarily that designated by CL. In this case, a new activation block is created on the *local stack* in order to save the current continuation:

```
; call/1 (+goal)
(defun |call| (x)
  (if (var? x)
      (let ((te (ultimate x PCE PCG)))
        (unless CP
                (if (<= BL CL) (setq L CL))
                (setq CP (CP CL) CL (CL CL)))
        (push_cont)
        (vset Mem (+ L 2) (cdr te))
        (vset Mem (+ L 3) Cut_pt)
        (setq CP (list (dec_goal (car te))) CL L)
        (maj_L 0))
    (push (dec_goal x) CP)))
(defmacro dec_goal (x)
  `(if (atom ,x) (list ,x) (cons (caar ,x) (cdr ,x))))
```

The activation block is created in a fashion analogous to the processing of the awakened goals (see section 13.4.4): its CP and CL fields save the current continuation, and its G field designates the execution environment (obviously global) of the goal to be proved. The current continuation registers are then updated. Finally, if `call` is performed in a last position, then it is first optimized (see section 13.3.2).

14.4. Arithmetic

The built-in predicates relevant to arithmetic are all based on the same model: they unify their third argument with the result of the operator applied to their two other arguments (which are expected to have a value at the time of the call):

```
; divi/3 (+int,+int,?int)
(defun |divi| (x y z) (uni z (floor (value1 x) (value1 y))))
; mod/3 (+int,+int,?int)
(defun |mod| (x y z) (uni z (rem (value1 x) (value1 y))))
; plus/3 (+int,+int,?int)
(defun |plus| (x y z) (uni z (+ (value1 x) (value1 y))))
; minus/3 (+int,+int,?int)
(defun |minus| (x y z) (uni z (- (value1 x) (value1 y))))
; mult/3 (+int,+int,?int)
(defun |mult| (x y z) (uni z (* (value1 x) (value1 y))))
```

For the comparisons, the test is performed by eventually triggering backtracking in case of failure:

```
; le/2 (+int,+int)
(defun |le| (x y) (if (> (value1 x) (value1 y)) 'fail))
; lt/2 (+int,+int)
(defun |lt| (x y) (if (>= (value1 x) (value1 y)) 'fail))
```

14.5. Type Tests and Conversion

These predicates determine the type of the term to which a variable is bound. They are all implemented by calling the corresponding Lisp boolean function:

```
; var/1 (?term)
(defun |var| (x) (unless (var? (value1 x)) 'fail))
; nonvar/1 (?term)
(defun |nonvar| (x) (if (var? (value1 x)) 'fail))
; atomic/1 (?term)
(defun |atomic| (x) (if (listp (value1 x)) 'fail))
; atom/1 (?term)
(defun |atom| (x) (unless (symbolp (value1 x)) 'fail))
; number/1 (?term)
(defun |number| (x) (unless (numberp (value1 x)) 'fail))
```

`name/2` has been chosen as an example of a predicate that performs a conversion between terms. It unifies the identifier with the list of ASCII codes that composes it. Since this predicate works both ways, it should be determined first which of the two arguments has a value at the time of the call (the case where the two arguments are bound is brought back to that of the first argument carrying a value):

```
; name/2 (?atom,?list)
(defun |name| (x y)
  (let ((b (value1 x)))
    (if (var? b)
        (uni x (impl (undo_1 (ultimate y PCE PCG))))
        (uni y (do_1 (expl b))))))
```

In order to build the identifier from the list of codes, the list should first be translated into a Lisp list, and then the `impl` primitive should be applied. In an analogous way, the corresponding list should be generated from the Lisp list of codes of the identifier obtained through `expl`:

```
(defun undo_1 (te)
  (let ((x (car te)) (e (cdr te)))
    (if (atom x)
        x
        (cons (undo_1 (val (cadr x) e)) (undo_1 (val (caddr x) e))))))
(defun do_1 (x)
  (if (atom x) x (list '(\. 2) (car x) (do_1 (cdr x)))))
(defun impl (l) (intern (map 'string #'int-char l)))
(defun expl (at) (map 'list #'char-int (string at)))
```

The `do_1` and `undo_1` primitives perform the conversion of lists between the Lisp and the Prolog representations (see section 11.1.1).

14.6. Extensions

Finally, let us focus on the predicates for the dynamic updating of the knowledge base (such as `asserta/1`, `assertz/1`, `clause/2`, or `retract/1`).

The implementation of such predicates does not cause any problem in an interpretive approach, aside from respecting the clause indexing. Consequently, each addition or deletion should take into account the initial organization in subsets (see section 11.1.3.).

The problem is more complex in the compiled versions, due both to the fundamental difference in the representation of the dynamic structures manipulated by the program and to the static coding of clauses set in their compiled format. Note that the implementation of dynamic addition (`asserta/1`, `assertz/1`, ...) does not cause any important problem. Indeed, it is sufficient to compile the extra clause and add it to the corresponding set, while respecting the indexing.

However, the implementation of the `clause/2` and `retract/1` predicates is trickier: the latter should be able to retrieve the initial structure of a clause from its compiled form. With respect to this matter, many approaches are feasible. The first approach consists simply in restricting the use of clauses additions and deletions to interpreted predicates. Another solution is to retain an image of the external form of a clause and to associate it with the compiled form [Clo85]. Finally, a third approach consists in recovering, through a decompilation process [Bue86], the initial image of a clause from its compiled form. For a complete description of this problem and its solutions, we strongly encourage the reader to refer to [Clo85] and [Bue86].

CONCLUSION

Research aimed at the design of machines dedicated to the execution of sequential Prolog has grown worldwide since the beginning of the 1980s. Among the various proposed designs, we retain the following two, which differ in their goals and thus in their approach toward the implementation of Prolog:

1. the PSI machine (Personal Sequential Inference) developed in the context of the Japanese fifth-generation project is specialized in the execution of Prolog. The PSI machine uses a classical memory management and architecture inspired by D. H. Warren's works.

2. the MALI machine (Machine Adaptée aux Langages Indéterministes), developed at the IRISA at Rennes, provides a genuine solution to the memory management problems through the use of a parallel real-time garbage-collector.

These two implementations will first be briefly presented. Each one will then be characterized according to the problems and solutions described in Part 2.

The PSI Machine

The most famous implementation is certainly the PSI machine developed as part of the fifth-generation project [NYY*83]. PSI is a single-user machine fully dedicated to the execution of Prolog [FF86].

The PSI machine language, KL0 (Kernel Language 0) is a subset of Prolog-Dec 10 and is directly executed by a microprogrammed interpreter. The ESP language (Extend Self-Contained Prolog), implemented on top of KL0, is a superset of an Object-Oriented Prolog. ESP is the programming language of the PSI. It was used in the design of SIMPOS, the PSI operating system. Two versions of the PSI machine have been successively implemented. The second version, PSI-2, is faster than PSI-1 and provides a better integration.

It is interesting to note that the implementation choices for the two versions of the PSI are based on D. H. Warren's proposals: PSI-1 is based on Prolog-Dec 10 and PSI-2 is based on Warren's Abstract Machine. We will not expand here on Warren's works: they have been abundantly covered in Part 2.

The MALI Machine

The MALI machine [BCRU84], [BCRU86] embeds three original ideas with respect to the different approaches that have been described.

First, MALI was designed with the purpose of defining a nondeterministic oriented languages machine, for which Prolog is only a particular case [CHR87]. Indeed, even though MALI uses a standard *or* control strategy, it does not fix an a priori strategy for the *and* control.

Moreover, MALI is conceived as a kit (processor and memory) that can be embedded in middle-sized hosts instead of being an autonomous independent workstation.

Finally, MALI has a unique memory management based on a parallel real-time garbage-collector. Indeed, whereas backtracking is implemented in the classical manner, the space allocated for resolvants and terms created by the forward process is not managed as a stack but rather is allocated in MALI's dedicated memory and recovered in parallel by a garbage-collector that uses Baker's algorithm [Bak78].

Overview of MALI

MALI owns a garbage-collector that fits logic programming [BCRU84], [BCRU86], [CHR87]. The machine is composed of:

1. a dedicated memory that allows for the representation of terms and resolvants states; and

2. a micro-programmed processor that carries two functions:

   a. a set of commands that performs the constructing and accessing of Prolog terms—the implementation of the forward process and the backtracking management;

   b. a real-time parallel memory recovery system.

MALI has been implemented in various hardware and software configurations [BCRU85]. Hardware versions are built as a module (processor and memory) that can be incorporated in a host microcomputer.

The tasks distribution and the communication between MALI and the host system is performed in the following manner [BCRU85]:

- The dedicated memory of MALI holds the representation of the dynamic structures (terms instances and resolvents), whereas the clauses and terms models (static structures) are coded in a separate memory space on the host system;

- The host processor, which controls the execution of Prolog, sequentially provides a set of commands that are to be executed by the MALI processor. Once a command is sent, the host waits for the result to be returned by MALI.

The Implementation of Prolog under MALI

The implementation of Prolog under MALI differs from the classical approach presented in Part 2 in four essential ways. First, MALI performs the duplication process without an environment allocation. Second, the *or* control management is hard-wired, and the resulting backtracking and restoring problems are handled automatically. Third, MALI does not assume a fixed *and* control strategy; this is left to the programmer. Fourth, space saving is achieved not through memory management (*local stack* and *global stack* with a garbage-collector) but through a unique parallel space-saving mechanism during the forward process.

Clause Duplication

MALI uses structure copying (see chapter 5). However, the duplication of a clause is achieved not in the classical way by associating it with an environment (see chapter 4) but instead by direct copying, in MALI memory, of the body of the selected clause in the head of the current resolvant. Thus, terms instances and Prolog resolvants are directly represented as MALI terms.

Duplication through an environment allocation is also feasible on MALI [CHR87]. This alternative has not been retained since the potential gain would be canceled by the complications it would bring at the implementation level [BCRU85].

The Implementation of the *Or* Control

The *or* control is directly managed by MALI, which features two internal stacks: the search stack and the *trail*. These implement backtracking. The classical architecture that solves the problems specific to backtracking has been retained: the search stack stores the choice points while the *trail* restores previous states of computation (see section 4.3).

However, since the decision has been made to represent resolvants as MALI terms, each rerun part holds two fields: a term associated with the resolvant status, and the status of the corresponding term on the *trail*. This representation of the choice points allows in particular the generalization of the Prolog *cut* through its association with a parameter that describes the status of the search stack. Finally, the saving of postbound variables is directly handled by the MALI binding mechanism.

The *And* Control Strategy

Unlike the fixed *or* control, MALI does not assume any kind of *and* control. This control is to be defined by the implementor. Accordingly, nonstandard strategies have been implemented under MALI, including the delaying mechanism associated with the *freeze/2* predicate [RH86].

Two solutions have been proposed. The first one is based on the explicit management of the sequences of frozen goals represented as lists. The other one, less flexible but more efficient, is based on the direct binding of the freezing variable with the delayed goal, according to the principle described in chapter 10.

Space Saving

The space saving on MALI is based not on a stacks updating mechanism (a *local stack* and a *global stack* with a garbage-collector) but rather on a proprietary collector that takes nondeterminism into account and makes use of the representation of resolvants as MALI terms. Garbage-collecting is triggered by a specific command, which makes its running in parallel with the forward process feasible. For Prolog, the optimal cycle seems to be the resolution step [CHR87].

In order to free the memory space that became useless for the rest of the proof, the MALI collector starts with the terms saved on the search stack and covers all the corresponding subterms. Each variable binding is then analyzed in order to determine if it is necessary to maintain it.

The particularity of the MALI collector is that it is able to detect the uselessness of a binding by comparing the timing of a binding with the information provided by the *trail*. If the binding value is determined to be useless, the variable is restored to its initial unbound state (early reset). If the variable is determined to be, every occurrence is replaced by its value (variable shunting). Therefore, the freeing of the memory space takes into account the characteristics of backtracking. We recommend [BCRU84], [BCRU86] for a detailed presentation on the collector's mechanism and its implementation, based on Baker's algorithm [Bak78].

MINI-CPROLOG

```
; mini-Cprolog
; boot.lsp
;

(defun mini-Cprolog ()                      ; to run it
  (banner)
  (format t "~A~%" (load "mlg.Start"))
  (myloop (read_prompt)))

(defun read_prompt ()
  (terpri)
  (format t "| ?- ")
  (read_code_tail))

(defun banner ()
  (dotimes (i 2) (terpri))
  (format t "Mini-Cprolog~%")
  (dotimes (i 2) (terpri)))

(defun l ()
  (format t "Back to Mini-Cprolog~%")       ; to re-enter from Lisp
  (myloop (read_prompt)))
```

```
; Mini-Cprolog
; parser.lsp
;

(defun read_code_cl ()                      ; to read and code a clause
  (let ((*lvarloc ()) (*lvarglob ()))
    (let ((x (read_clause (rchnsep))))
      (cons (cons (length *lvarloc)          ; number of local vars
                  (length *lvarglob))        ; number of global vars
            (c x)))))

(defun read_code_tail ()                     ; idem for a query
```

```
    (setq *lvarloc () *lvarglob ())
    (let ((x (read_tail (rchnsep))))
      (cons (cons (length *lvarloc) (length *lvarglob)) (c x))))

(defun read_pred (ch)                            ; to read a literal
  (let ((nom (read_atom ch)) (c (rchnsep)))
    (if (char= c #\()
        (cons nom (read_args (rchnsep) 1))
      (progn (unread-char c) (list nom)))))
```

```
; mini-Cprolog
; blocks.lsp
;

; I. Registers
;
(defconstant BottomG 1)                     ; bottom of global stack
(defconstant BottomL 3000)                  ; bottom of local stack
(defconstant BottomTR 8000)                 ; bottom of trail
(defconstant TopTR 10000)                   ; max of trail

(defvar Mem (make-array 10000 :initial-element 0))
(defvar TR)                                 ; top of trail
(defvar L)                                  ; top of local stack
(defvar G)                                  ; top of global stack
(defvar CP)                                 ; current continuation
(defvar CL)
(defvar BL)                                 ; last choice point
(defvar BG)
(defvar PC)                                 ; current goal and its
(defvar PCE)                                ; environments
(defvar PCG)
(defvar Duboulot)

(defmacro vset (v i x) '(setf (svref ,v ,i) ,x))

; II. Local Stack
; deterministic block : [CL CP E G]
;
(defmacro CL (b) '(svref Mem ,b))
(defmacro CP (b) '(svref Mem (1+ ,b)))
(defmacro G (b) '(svref Mem(+ ,b 2)))
(defmacro E (b) '(+ ,b 3))
```

```lisp
(defmacro push_cont ()                      ; saves continuation
  '(progn (vset Mem L CL) (vset Mem (1+ L) CP)))

(defmacro push_E (n)                        ; allocates local env.
  '(let ((top (+ L 3 ,n)))
     (if (>= top BottomTR)
         (throw 'debord (print "Local Stack Overflow")))
     (dotimes (i ,n top)                    ; n local free variables
       (vset Mem (decf top) (cons 'LIBRE BottomG)))))

(defmacro maj_L (nl)                        ; updates top of local stack
  '(incf L (+ 3 ,nl)))

; choice block : [TR BP BL]
;
(defmacro BL (b) '(svref Mem (1- ,b)))
(defmacro BP (b) '(svref Mem (- ,b 2)))
(defmacro TR (b) '(svref Mem (- ,b 3)))

(defmacro push_choix ()                     ; allocates a rerun part
  '(progn (vset Mem L TR)                   ; current top of trail
          (vset Mem (incf L 2) BL)          ; saves last choice block
          (setq BL (incf L) BG G)))         ; new last choice block

(defmacro push_bpr (reste)                  ; subset of clauses still
  '(vset Mem (- BL 2) ,reste))              ; to be considered

(defmacro pop_choix ()
  '(setq L (- L 3)                          ; pops the block
         BL (BL BL)                         ; updates backtrack
         BG (if (zerop BL) 0 (G BL))))      ; specific registers

; III. Global Stack
;
(defmacro push_G (n)                        ; allocates a global env
  '(let ((top (+ G ,n)))
     (if (>= top BottomL)
         (throw 'debord (print "Global Stack Overflow")))
     (dotimes (i ,n (vset Mem (+ L 2) G))   ; field G of local stack
       (vset Mem (decf top) (cons 'LIBRE BottomG)))))

(defmacro maj_G (ng) '(incf G ,ng))

; IV. Trail
;
(defmacro pushtrail (x)                     ; remembers that x
  '(progn                                   ; is post-bound
     (if (>= TR TopTR) (throw 'debord (print "Trail Overflow")))
     (vset Mem TR ,x)
     (incf TR)))
```

```lisp
(defmacro poptrail (top)                    ; pops the trail until top
  '(do () ((= TR ,top))                     ; restore vars to free state
     (vset Mem (svref Mem (decf TR)) (cons 'LIBRE BottomG))))
```

```lisp
; mini-Cprolog
; unify.lsp
;

(defmacro bindte (x sq e)                   ; binds var x to
  '(progn                                   ; the pair (sq,e)
     (if (or (and (> ,x BottomL)            ; local var post-bound?
                  (< ,x BL))
             (< ,x BG))                     ; global var post-bound ?
         (pushtrail ,x))                    ; push it on the trail
     (rplaca (svref Mem ,x) ,sq)            ; binds it
     (rplacd (svref Mem ,x) ,e)))

(defun unify (t1 el1 eg1 t2 el2 eg2)        ; unifies 2 lists of args
  (catch 'impossible
    (do () ((null t1))
      (unif (ultimate (pop t1) el1 eg1)
            (ultimate (pop t2) el2 eg2)))))
```

```lisp
; Mini-CProlog
; resol.lsp
;

(defun forward ()                           ; forward process
  (do () ((null Duboulot) (format t "no More ~%"))
    (cond
      ((null CP) (answer))                  ; empty resolvent
      ((load_PC)                            ; selects the first goal
       (cond
         ((user? PC)                        ; user defined
          (let ((d (def_of PC)))            ; associated set of clauses
            (if d (pr d) (backtrack))))
         ((builtin? PC)                     ; builtin predicate
```

```
                (if (eq (apply (car PC) (cdr PC)) 'fail)
                    (backtrack)                   ; evaluation fails
                    (cont_det)))                  ; else new continuation
                ((backtrack)))))))                ; undefined predicate

(defun load_PC ()                            ;first goal and its env
  (setq PC (car CP) PCE (E CL) PCG (G CL)))

(defun pr (paq)                          ; if more than one clause
  (if (cdr paq)                          ; allocates a rerun part
      (progn (push_choix) (pr_choice paq))
    (pr_det (car paq))))

(defun pr_det (c)                            ; only one clause candidate
  (if (eq (unify (largs PC) PCE PCG
                 (largs (head c)) (push_E (nloc c)) (push_G (nglob c)))
          'fail)                           ; tries to unify
      (backtrack)                          ; failure
    (progn                                 ; success
      (if (tail c)                         ; for a rule
          (progn (push_cont) (setq CP (tail c) CL L) (maj_L (nloc c)))
        (cont_det))                        ; for a fact
      (maj_G (nglob c)))))

(defun cont_det ()                           ; looks for next continua.
  (if (cdr CP)                               ; and siblings ?
      (setq CP (cdr CP))                     ; done
    (progn                                   ; no intermediate choice pts
      (if (< BL CL) (setq L CL))             ; updates the local stack
      (do ()                                 ; searching next cont.
          ( (or (cdr (CP CL)) (zerop (CL CL)))
            (setq CP (cdr (CP CL)) CL (CL CL)))
        (setq CL (CL CL))
        (if (< BL CL) (setq L CL))))))

(defun pr_choice (paq)                       ; more than one candidate
  (let* ((resu (shallow_backtrack paq)) ; determines first success
         (c (car resu))
         (r (cdr resu)))
    (cond ((null r)                          ; only one candidate remains
           (pop_choix)                       ; pops the rerun part
           (pr_det c))                       ; proof is now deterministic
          ( (push_bpr r)                     ; else updates the field BP
            (push_cont)                      ; saves current continuation
            (if (tail c)                     ; looking for the new one
                (setq CP (tail c) CL L)
              (if (cdr CP)
                  (setq CP (cdr CP))
                (do ()
                    ((or (cdr (CP CL)) (zerop (CL CL)))
                     (setq CP (cdr (CP CL)) CL (CL CL)))
                  (setq CL (CL CL)))))))
```

```
            (maj_L (nloc c))                ; terminates local block
            (maj_G (nglob c))))))           ; terminates global block

(defun shallow_backtrack (paq)             ; looks for a first success
  (if (and (cdr paq)                        ; more than one candidate
           (eq (unify (largs PC)            ; and unification fails
                      PCE
                      PCG
                      (largs (head (car paq)))
                      (push_E (nloc (car paq)))
                      (push_G (nglob (car paq))))
               'fail))
      (progn (poptrail (TR BL))             ; restores the environments
             (shallow_backtrack (cdr paq))) ; again with next clause
    paq))

(defun backtrack ()
  (if (zerop BL)                            ; no more choice points
      (setq Duboulot nil)                   ; neds the proof
    (progn                                  ; updates backtracking
     (setq L BL G BG                        ; specific registers
           PC (car (CP L))                  ; restore the goal and
           PCE (E (CL L)) PCG (G (CL L))    ; its environments
           CP (CP L) CL (CL L))
     (poptrail (TR BL))                     ; restores the environments
     (pr_choice (BP L)))))                  ; restarts the proof

(defun myloop (c)                          ; top level loop
  (setq G BottomG L BottomL TR BottomTR    ; init registers
        CP nil CL 0  BL 0 BG BottomG Duboulot t)
  (push_cont)                              ; saves empty continuation
  (push_E (nloc c))                        ; allocates env for query
  (push_G (nglob c))
  (setq CP (cdr c) CL L)                   ; current continuation
  (maj_L (nloc c))                         ; terminates block
  (maj_G (nglob c)) (read-char)
  (catch 'debord (forward))                ; to catch stacks overflow
  (myloop (read_prompt)))
```

```
; Mini-CProlog
; pred.lsp
;

(defvar Ob_Micro_Log                       ; list of builtins
  '(|write| |nl| |tab| |read| |get| |get0|
```

```
              |var| |nonvar| |atomic| |atom| |number|
              ! |fail| |true|
              |divi| |mod| |plus| |minus| |mult| |le| |lt|
              |name| |consult| |abolish| |cputime| |statistics|))

(mapc                                    ; tags the builtins
 #' (lambda (x) (setf (get x 'evaluable) t))
 Ob_Micro_Log)

;cut/0
(defun ! (n)
  (unless (< BL CL)
          (do () ((< BL CL)) (setq BL (BL BL)))
          (setq BG (if (zerop BL) BottomG (G BL)) L (+ CL 3 n))))

; statistics/0
(defun |statistics| ()
  (format t " local stack : ~A (~A used)~%"
          (- BottomTR BottomL) (- L BottomL))
  (format t " global stack : ~A (~A used)~%"
          BottomL (- G BottomG))
  (format t " trail : ~A (~A used)~%"
          (- TopTR BottomTR) (- TR BottomTR)))
```

```
% Startup Mini-Cprolog
%

    $ repeat.
    $ repeat :- repeat.

    $ eq(X,X).
    $ neq(X,X):- !, fail.
    $ neq(_1,_2).

    $ conc([],X,X) .
    $ conc([T|Q],L,[T|R]) :- conc(Q,L,R).
    $ member(X,[X|_1]).
    $ member(X,[_1|Z]) :- member(X,Z).
    $ del(X,[X|Y],Y).
    $ del(X,[Y|Z],[Y|R]) :- del(X,Z,R).

    $ wb(X) :- write(X), write(' ').
    $ wf(X) :- write(X), nl, fail.
```

APPENDIX B

MINI-WAM

```
; mini-Wam
; boot.lsp
;

(defun mini-Wam ()
  (banner)
  (format t "~A~%" (load "mlg.Start"))
  (myloop (read_prompt)))

(defun read_prompt ()
  (terpri)
  (format t "| ?- ")
  (read_code_tail))

(defun banner ()
  (dotimes (i 2) (terpri))
  (format t "mini-Wam~%")
  (dotimes (i 2) (terpri)))

(defun l ()
  (format t "Back to mini-Wam top-level~%")
  (myloop (read_prompt)))
```

```
; mini-Wam
; parser.lsp
;

(defvar *lvar nil)                        ; list of vars
(set-macro-character #\% (get-macro-character #\;))

(defun rch ()                             ; skips Newline
  (do ((ch (read-char) (read-char)))
      ((char/= ch #\Newline) ch)))
(defun rchnsep ()                         ; skips Newline and Space
  (do ((ch (rch) (rch)))
      ((char/= ch #\space) ch)))
```

```
(defun special (ch) (char= ch #\_))
(defun alphanum (ch) (or (alphanumericp ch) (special ch)))
(defun valdigit (ch) (digit-char-p ch))

(defun read_number (ch)                  ; next integer
  (do ((v (valdigit ch) (+ (* v 10) (valdigit (read-char)))))
      ((not (digit-char-p (peek-char))) v)))

(defun implode (lch) (intern (map 'string #'identity lch)))

(defun read_atom (ch)                     ; next normal atom
  (do ((lch (list ch) (push (read-char) lch)))
      ((not (alphanum (peek-char))) (implode (reverse lch)))))

(defun read_at (ch)                       ; next special atom
  (do ((lch (list ch) (push (read-char) lch)))
      ((char= (peek-char) #\') (read-char) (implode (reverse lch)))))

(defun read_string (ch)                   ; Prolog list of Ascii codes
  (do ((lch (list (char-int ch)) (push (char-int (read-char)) lch)))
      ((char= (peek-char) #\") (read-char) (do_l (reverse lch)))))

(defun read_var (ch)                      ; next variable
  (let ((v (read_atom ch)))
    (cons 'V
          (position v (if (member v *lvar)
                          *lvar
                          (setq *lvar (append *lvar (list v))))))))

(defun read_simple (ch)                   ; next simple term
  (cond
    ((or (special ch) (upper-case-p ch)) (read_var ch))
    ((digit-char-p ch) (read_number ch))
    ((char= ch #\") (read_string (read-char)))
    ((char= ch #\') (read_at (read-char)))
    (t (read_atom ch))))

(defun read_fct (ch)                      ; next functional term
  (let ((fct (read_simple ch)) (c (rchnsep)))
    (if (char= c #\()                     ; reads its argument
        (let ((la (read_args (rchnsep))))
          (cons (list fct (length la)) la)) ; adds its descriptor
      (progn (unread-char c) fct))))

(defun read_args (ch)                     ; args of a functional term
  (let ((arg (read_term ch)))
    (if (char= (rchnsep) #\,)
        (cons arg (read_args (rchnsep)))
      (list arg))))
```

```
(defun read_list (ch)                    ; next list
  (if (char= ch #\])
      ()                                 ; empty list
    (let ((te (read_term ch)))           ; gets the head
      (case (rchnsep)
            (#\, (list '(\. 2) te (read_list (rchnsep))))
            (#\| (prog1                  ; dotted pair
                    (list '(\. 2) te (read_term (rchnsep)))
                  (rchnsep)))
            (#\] (list '(\. 2) te nil))))))

(defun read_term (ch)                    ; next term
  (if (char= ch #\[) (read_list (rchnsep)) (read_fct ch)))

(defun read_tail (ch)                    ; reads the body of a clause
  (let ((tete (read_pred ch)))
    (if (char= (rchnsep) #\.)
        (list tete)                      ; first goal
      (cons tete (read_tail (rchnsep))))))

(defun read_clause (ch)                  ; reads a clause
  (let ((tete (read_pred ch)))           ; gets the head
    (if (char= (rchnsep) #\.)
        (list tete)                      ; a fact
      (progn (read-char)                 ; else a rule
             (cons tete (read_tail (rchnsep)))))))

(defun read_code_cl ()                   ; reads and codes a clause
  (let ((*lvar ()))
    (let ((x (read_clause (rchnsep))))
      (cons (length *lvar) (c x)))))

(defun read_code_tail ()                 ; reads a query
  (setq *lvar ())
  (let ((x (read_tail (rchnsep))))
    (cons (length *lvar) (append (c x) (list '(|true|))))))

(defun c (l)
  (if l (cons (if (eq (caar l) '!)
                  (list '! (length *lvar)) ; the cut and env size
                (car l))
              (c (cdr l)))))

(defun read_pred (ch)                    ; reads a predicate
  (let ((nom (read_atom ch)) (c (rchnsep)))
    (if (char= c #\()
        (cons nom (read_args (rchnsep)))
      (progn (unread-char c) (list nom)))))
```

```
; mini-Wam
; blocks.lsp
;

; I. Registers
;

(defconstant BottomG 0)                 ; bottom of the heap
(defconstant BottomL 2000)              ; bottom of local stack
(defconstant BottomTR 8000)             ; bottom of trail
(defconstant A 10000)                   ; registers Ai

(defvar Mem (make-array 10050 :initial-element 0))
(defvar TR)                             ; top of trail
(defvar L)                              ; top of local stack
(defvar G)                              ; top of heap
(defvar CP)                             ; current continuation
(defvar CL)
(defvar Cut_pt)                         ; specific cut
(defvar BL)                             ; last choice point
(defvar BG)
(defvar PC)                             ; current goal and its
(defvar PCE)                            ; environment
(defvar Duboulot)

(defmacro vset (v i x) '(setf (svref ,v ,i) ,x))
(defmacro tcons (des largs)             ; tags instances
  '(cons (append ,des (list '#\*)) ,largs))
(defmacro des (te) '(car ,te))
(defmacro arity (des) '(cadr ,des))
(defmacro functor (des) '(car ,des))
(defmacro largs (x) '(cdr ,x))
(defmacro var? (x) '(and (consp ,x) (numberp (cdr ,x))))
(defmacro list? (x) '(eq (functor (des ,x)) '\.))

; II. Local Stack
;

;deterministic block [CL CP Cut E]
;
(defmacro CL (b) '(svref Mem ,b))
(defmacro CP (b) '(svref Mem (1+ ,b)))
(defmacro Cut (b) '(svref Mem (+ ,b 2)))
(defmacro E (b) '(+ ,b 3))

(defmacro push_cont ()                  ; saves continuation
  '(progn (vset Mem L CL) (vset Mem (1+ L) CP)))

(defmacro push_E (n)                    ; allocates env of size n
  '(let ((top (+ L 3 ,n)))
     (if (>= top BottomTR)
         (throw 'debord (print "Local Stack Overflow")))
```

```
      (vset Mem (+ L 2) Cut_pt)              ; saves cut point
      (dotimes (i ,n top) (vset Mem (decf top) (cons 'V top)))))

(defmacro maj_L (nl) '(incf L (+ 3 ,nl)))


;Rerun parts : [a1 .. an A BCP BCL BG BL BP TR]
;
(defmacro TR (b) '(svref Mem (1- ,b)))
(defmacro BP (b) '(svref Mem (- ,b 2)))
(defmacro BL (b) '(svref Mem (- ,b 3)))
(defmacro BG (b) '(svref Mem (- ,b 4)))
(defmacro BCL (b) '(svref Mem (- ,b 5)))
(defmacro BCP (b) '(svref Mem (- ,b 6)))
(defmacro A (b) '(svref Mem (- ,b 7)))

(defun save_args ()                         ; save args of current goal
  (dotimes (i (svref Mem A) (vset Mem (incf L i) i))
           (vset Mem (+ L i) (svref Mem (+ A i 1)))))

(defun push_choix ()                        ; allocates a choice point
  (save_args)                               ; goal args
  (vset Mem (incf L) CP)                     ; continuation
  (vset Mem (incf L) CL)
  (vset Mem (incf L) G)                      ; top of heap
  (vset Mem (incf L) BL)                     ; last choice point
  (vset Mem (incf L 2) TR)                   ; top of trail
  (setq BL (incf L) BG G))                   ; new last choice point

(defun push_bpr (reste) (vset Mem (- BL 2) reste))
(defmacro size_C (b) '(+ 7 (A ,b)))

(defun pop_choix ()
  (setq L (- BL (size_C BL))                 ; pops the block
        BL (BL BL)                           ; updates backtrack reg.
        BG (if (zerop BL) BottomG (BG BL))))

; III. Copy Stack (Heap)
;
(defmacro push_G (x)                         ; to push a term instance
  '(if (>= (incf G) BottomL)
       (throw 'debord (print "Heap Overflow"))
     (vset Mem G ,x)))

(defmacro adr (v e) '(+ (cdr ,v) ,e))

(defun copy (x e)                            ; copy skeleton x in
  (cond                                      ; environment e
   ((var? x)
    (let ((te (ult (adr x e))))              ; dereferencing
      (if (var? te) (genvar (cdr te)) te)))  ; a new one
```

```
    ((atom x) x)
    ((tcons (des x) (cop (largs x) e)))))  ; recursively

(defun cop (l e)                           ; copy while tagging
  (if l (cons (copy (car l) e) (cop (cdr l) e))))

(defmacro recopy (x e) '(push_G (copy ,x ,e)))

(defmacro copy? (te) '(cddr (des ,te)))  ; instance ?

;IV. Trail
;

(defmacro pushtrail (x)
  '(cond ((>= TR A) (throw 'debord (print "Trail Overflow")))
         ((vset Mem TR ,x) (incf TR))))

(defmacro poptrail (top)
  '(do () ((= TR ,top))
      (let ((v (svref Mem (decf TR)) )) (vset Mem v (cons 'V v)))))
```

```
; mini-Wam
; utili.lsp
;

(defmacro nvar (c) '(car ,c))              ; number of vars
(defmacro head (c) '(cadr ,c))             ; head of a clause
(defmacro tail (c) '(cddr ,c))             ; body of a rule
(defmacro pred (g) '(car ,g))              ; predicate symbol

(defmacro user? (g)                        ; user defined predicate
  '(get (pred ,g) 'def))
(defmacro builtin? (g)                     ; builtin predicate
  '(get (pred ,g) 'evaluable))
(defmacro def_of (g)                       ; gets the subset of clauses
  '(get (pred ,g)
        (if (largs ,g) (nature (ultimate (car (largs ,g)) PCE)) 'def)))

(defun nature (te)                         ; determines the associated
  (cond                                    ; subset according to
    ((var? te) 'def)                       ; clause indexing
    ((null te) 'empty)
    ((atom te) 'atom)
    ((list? te) 'list)
    (t 'fonct)))
```

```
(defun add_cl (pred c ind)              ; adds a clause to the end
   (setf (get pred ind) (append (get pred ind) (list c))))

(set-macro-character                    ; to automatically read,
 #\$                                    ; code and index
 #'(lambda (stream char)
     (let* ( (*standard-input* stream) (c (read_code_cl)))
       (add_cl (pred (head c)) c 'def)  ; always in the var subset
       (if (largs (head c))
           (let ((b (nature (car (largs (head c))))))
             (if (eq b 'def)            ; first arg is a var
                 (mapc                  ; add c to all subsets
                  #' (lambda (x) (add_cl (pred (head c)) c x))
                   '(atom empty list fonct))
                 (add_cl (pred (head c)) c b))))) ; add to only one
     (values)))

(defun answer ()                        ; the answer to a query
  (printvar)                            ; resulting bindings
  (if (zerop BL)                        ; no more choice points
      (setq Duboulot nil)
      (if (and (princ "More : ")        ; asks the user
               (string= (read-line) ";")) ; if necessary
          (backtrack)                   ; forces backtracking
          (setq Duboulot nil))))

(defun printvar ()                      ; prints the values of vars
  (if (null *lvar)                      ; ground query
      (format t "Yes ~%")
      (let ((n -1))
        (mapc                           ; proceeds each variable
         #' (lambda (x)
              (format t "~A = " x)
              (write1 (ult (+ (incf n) (E BottomL)))) (terpri))
         *lvar))))
```

```
; mini-Wam
; unify.lsp
;
(defun ult (n)                          ; dereferences a var
  (let ((te (svref Mem n)))
    (if (and (var? te) (/= (cdr te) n)) (ult (cdr te)) te)))
```

```lisp
(defun ultimate (x e) (if (var? x) (ult (adr x e)) x))
(defun val (x) (if (var? x) (ult (cdr x)) x))

(defmacro bind (x te)              ; binds x to instance te
  '(progn     `                    ; if post-bound
     (if (or (and (> ,x BottomL) (< ,x BL)) (<= ,x BG))
         (pushtrail ,x))           ; trail it
     (vset Mem ,x ,te)))
(defun bindte (xadr y)             ; atom or instance
  (if (or (atom y) (copy? y))
      (bind xadr y)                ; direct
    (bind xadr (recopy y (E L))))) ; else constructing
(defun genvar (x) (bind x (push_G (cons 'V G))))
(defmacro bindv (x y)              ; L2 binding
  '(if (< (cdr ,x) (cdr ,y)) (bind (cdr ,y) ,x) (bind (cdr ,x) ,y)))

(defun unify_with (largs e)        ; unify head of clause
  (catch 'impossible               ; with registers Ai
    (do ((i (1+ A) (1+ i)))
        ((null largs))
        (unif (svref Mem i) (ultimate (pop largs) e)))))

(defun unif (x y)                  ; unify two terms
  (cond
   ((eql x y) t)
   ((var? y) (if (var? x)
                 (if (= (cdr x) (cdr y)) t (bindv y x))
               (bindte (cdr y) x)))
   ((var? x) (bindte (cdr x) y))
   ((or (atom x) (atom y)) (throw 'impossible 'fail))
   ((let ((b (copy? y)) (dx (pop x)) (dy (pop y)))
      (if (and (eq (functor dx) (functor dy))
               (= (arity dx) (arity dy)))
          (do () ((null x))
              (unif (val (pop x))
                    (if b (val (pop y)) (ultimate (pop y) (E L)))))
        (throw 'impossible 'fail))))))
```

```lisp
; mini-Wam
; resol.lsp
;

(defun forward ()
  (do () ((null Duboulot) (format t "no More ~%"))
      (cond ((null CP) (answer))   ; empty resolvent
```

```
        ( (load_PC)                    ; selects first goal
          (cond
            ((user? PC)                 ; user defined
             (let ((d (def_of PC)))    ; associated set of clauses
               (if d (pr2 d) (backtrack))))
            ((builtin? PC)              ; builtin
             (if (eq (apply (car PC) (cdr PC)) 'fail)
                 (backtrack)            ; evaluation fails
                 (cont_eval)))          ; else new continuation
            ((backtrack)))))))          ; undefined predicate

(defun load_PC ()
  (setq PC (pop CP) PCE (E CL) Cut_pt BL))

(defun load_A (largs el)               ; loads A1 registers with
  (dotimes (i (length largs) (vset Mem A i)) ; goal's args
         (vset Mem
               (+ A i 1)
               (let ((te (ultimate (pop largs) el)))
                 (cond
                   ((atom te) te)
                   ((var? te) (if (unsafe? te) (genvar (cdr te)) te))
                   ((copy? te) te)
                   ((recopy te el)))))))

(defun unsafe? (x)                     ; last call and current env
  (and (not CP) (>= (cdr x) CL)))

(defun pr2 (paq)                       ; proves PC
  (load_A (largs PC) PCE)              ; loads its arguments
  (if CP
      (pr paq)
    (progn                             ; if last call and no interm.
      (if (<= BL CL) (setq L CL))      ; choice points then LCO
      (setq CP (CP CL) CL (CL CL))     ; next continuation
      (pr paq))))

(defun cont_eval ()
  (unless CP
         (if (<= BL CL) (setq L CL)) (setq CP (CP CL) CL (CL CL))))

(defun pr (paq)
  (if (cdr paq)                        ; alloc choice point
      (progn (push_choix) (pr_choice paq))
    (pr_det (car paq))))               ; else deterministic

(defun pr_det (c)
  (if (eq (unify_with (largs (head c)) (push_E (nvar c))) 'fail)
      (backtrack)
    (when (tail c)                     ; c is a rule
          (push_cont)                  ; saves current cont.
```

```
            (setq CP (tail c) CL L)        ; new one
            (maj_L (nvar c)))))             ; terminates local block

(defun pr_choice (paq)
  (let* ((resu (shallow_backtrack paq)) (c (car resu)) (r (cdr resu)))
    (cond ((null r)                        ; only one candidate remains
           (pop_choix)                     ; pops the rerun part
           (pr_det c))                     ; proof is now deterministic
          ( (push_bpr r)
            (when (tail c)                 ; c is a rule
                  (push_cont)              ; saves current cont.
                  (setq CP (tail c) CL L)  ; new one
                  (maj_L (nvar c)))))))    ; terminates local block

(defun shallow_backtrack (paq)             ; looks for a first success
  (if (and (cdr paq)                       ; more than one candidate
           (eq (unify_with                 ; and unification fails
                (largs (head (car paq)))
                (push_E (nvar (car paq)))) 'fail))
      (progn
        (poptrail (TR BL))                 ; restores env
        (setq G BG)
        (shallow_backtrack (cdr paq)))
    paq))

(defun backtrack ()
  (if (zerop BL)                           ; no more choice points
      (setq Duboulot nil)
    (progn                                 ; restores registers
      (setq L BL G BG Cut_pt (BL BL) CP (BCP L) CL (BCL L))
      (load_A2)
      (poptrail (TR BL))                   ; restores env.
      (pr_choice (BP L)))))                ; restart the proof

(defun load_A2 ()                          ; restores Ai registers
  (let ((deb (- L (size_C L))))
    (dotimes (i (A L) (vset Mem A i))
             (vset Mem (+ A i 1) (svref Mem (+ deb i)))))))

(defun myloop (c)
  (setq G BottomG L BottomL TR BottomTR ; inits registers
        CP nil CL 0 BL 0 BG BottomG Duboulot t Cut_pt 0)
  (push_cont)                              ; initial continuation
  (push_E (nvar c))                        ; env. for the query
  (setq CP (cdr c) CL L)                   ; current continuation
  (maj_L (nvar c)) (read-char)             ; ends block
  (catch 'debord (forward))
  (myloop (read_prompt)))
```

```
; mini-Wam
; pred.lsp
;

(defvar Ob_Micro_Log
      '(|write| |nl| |tab| |read| |get| |get0|
        |var| |nonvar| |atomic| |atom| |number|
        ! |fail| |true|
        |divi| |mod| |plus| |minus| |mult| |le| |lt|
        |name| |consult| |abolish| |cputime| |statistics|))
(mapc #' (lambda (x) (setf (get x 'evaluable) t)) Ob_Micro_Log)

(defmacro value (x)
    '(if (or (var? ,x) (atom ,x)) (ultimate ,x PCE) (copy ,x PCE)))
(defun uni (x y) (catch 'impossible (unif (value x) y)))

;write/1 (?term)
 (defun |write| (x) (write1 (value x)))
    (defun write1 (x)
      (cond
       ((null x) (format t "[]"))
       ((atom x) (format t "~A" x))
       ((var? x) (format t "X~A" (cdr x)))
       ((list? x) (format t "[")
        (writesl (val (cadr x)) (val (caddr x)))
        (format t "]"))
       ((writesf (functor (des x)) (largs x)))))
    (defun writesl (tete q)
      (write1 tete)
      (cond
       ((null q))
       ((var? q) (format t "|X~A" (cdr q)))
       (t (format t ",") (writesl (val (cadr q)) (val (caddr q))))))
    (defun writesf (fct largs)
      (format t "~A(" fct)
      (write1 (val (car largs)))
      (mapc #' (lambda (x) (format t ",") (write1 (val x))) (cdr largs))
      (format t ")"))

;nl/0
      (defun |nl| () (terpri))
;tab/1 (+int)
      (defun |tab| (x) (dotimes (i (value x)) (format t " ")))
;read/1 (?term)
      (defun |read| (x)
        (let ((te (read_terme)))
          (catch 'impossible
            (unif (value x) (recopy (cdr te) (push_E (car te)))))))

        (defun read_terme ()
          (let ((*lvar nil))
```

```
                (let ((te (read_term (rchnsep))))
                  (rchnsep) (cons (length *lvar) te))))
;get/1 (?car)
        (defun |get| (x) (uni x (char-int (rchnsep)))))
;get0/1 (?car)
        (defun |get0| (x) (uni x (char-int (read-char)))))


;var/1 (?term)
        (defun |var| (x) (unless (var? (value x)) 'fail))
;nonvar/1 (?term)
        (defun |nonvar| (x) (if (var? (value x)) 'fail))
;atomic/1 (?term)
        (defun |atomic| (x) (if (listp (value x)) 'fail))
;atom/1 (?term)
        (defun |atom| (x) (unless (symbolp (value x)) 'fail))
;number/1 (?term)
        (defun |number| (x) (unless (numberp (value x)) 'fail))

;cut/0
        (defun ! (n)
          (setq BL (Cut CL)
                BG (if (zerop BL) BottomG (BG BL))
                L (+ CL 3 n)))               ; updates local stack
;fail/0
        (defun |fail| () 'fail)
;true/0
        (defun |true| ())

;divi/3 (+int,+int,?int)
        (defun |divi| (x y z) (uni z (floor (value x) (value y))))
;mod/3 (+int,+int,?int)
        (defun |mod| (x y z) (uni z (rem (value x) (value y))))
;plus/3 (+int,+int,?int)
        (defun |plus| (x y z) (uni z (+ (value x) (value y))))
;minus/3 (+int,+int,?int)
        (defun |minus| (x y z) (uni z (- (value x) (value y))))
;mult/3 (+int,+int,?int)
        (defun |mult| (x y z) (uni z (* (value x) (value y))))
;le/2 (+int,+int)
        (defun |le| (x y) (if (> (value x) (value y)) 'fail))
;lt/2 (+int,+int)
        (defun |lt| (x y) (if (>= (value x) (value y)) 'fail))

;name/2 (?atom,?list)
(defun |name| (x y)
  (let ((b (value x)))
     (if (var? b)
          (uni x (impl (undo_l (value y))))
          (uni y (do_l (expl b))))))
(defun undo_l (x)
  (if (atom x)
```

```
      x
      (cons (undo_l (val (cadr x))) (undo_l (val (caddr x)))))))
(defun do_l (x)
  (if (atom x) x (list '(\. 2) (car x) (do_l (cdr x)))))
(defun impl (l) (intern (map 'string #'int-char l)))
(defun expl (at) (map 'list #'char-int (string at)))

 ;consult/1 (+atom)
      (defun |consult| (f) (format t "~A~%" (load (value f))))
; abolish/1 (+atom)
(defun |abolish| (p)
  (mapc #'(lambda (x) (setf (get p x) nil))
        '(atom empty list fonct def)))

; cputime/1 (?int)
(defun |cputime| (x)
  (uni x (float (/ (get-internal-run-time)
                   internal-time-units-per-second)))))

; statistics/0
(defun |statistics| ()
  (format t " local stack : ~A (~A used)~%"
          (- BottomTR BottomL) (- L BottomL))
  (format t " global stack : ~A (~A used)~%"
          BottomL (- G BottomG))
  (format t " trail : ~A (~A used)~%"
          (- A BottomTR) (- TR BottomTR)))
```

```
% Startup MiniWam
%

    $ repeat.
    $ repeat :- repeat.

    $ eq(X,X).
    $ neq(X,X):- !, fail.
    $ neq(_1,_2).

    $ conc([],X,X) .
    $ conc([T|Q],L,[T|R]) :- conc(Q,L,R).
    $ member(X,[X|_1]).
    $ member(X,[_1|Z]) :- member(X,Z).
    $ del(X,[X|Y],Y).
    $ del(X,[Y|Z],[Y|R]) :- del(X,Z,R).

    $ wb(X) :- write(X), write(' ').
    $ wf(X) :- write(X), nl, fail.
```

APPENDIX C

MINI-PROLOG-II

```
; Mini-PrologII
; boot.lsp
;

(defun mini-PrologII ()                    ; to run it
  (banner)
  (format t "~A~%" (load "mlg.Start"))
  (myloop (read_prompt)))

(defun read_prompt ()
  (terpri)
  (format t "| ?- ")
  (read_code_tail))

(defun banner ()
  (dotimes (i 2) (terpri))
  (format t "Mini-PrologII~%")
  (dotimes (i 2) (terpri)))

(defun l ()
  (format t "Back to Mini-PrologII top-level~%")
  (myloop (read_prompt)))
```

```
; mini-PrologII
; parser.lsp
;

(defun read_code_cl ()                     ; to read and code a clause
  (let ((*lvarloc ()) (*lvarglob ()))
    (let ((x (read_clause (rchnsep))))
      (maj_locglob (car x) (car (last x)))
      (cons                                ; number of local and global
       (cons (length *lvarloc) (length *lvarglob)) ; vars
       (c x)))))
```

```
(defun read_code_tail ()                    ; idem for a query
  (setq *lvarloc () *lvarglob ())
  (let ((x (read_tail (rchnsep))))
    (cons
     (cons (length *lvarloc) (length *lvarglob))
     (append (c x) (list '(|true|))))))

(defun read_pred (ch)                       ; to read a literal
  (let ((nom (read_atom ch)) (c (rchnsep)))
    (if (char= c #\()
        (cons nom
              (read_args (rchnsep)
                         (if (member nom '(|dif| |freeze|))
                             2              ; a var occurring in a dif
                             1)))           ; or in a freeze is global
      (progn (unread-char c) (list nom)))))

(defun unsafe? (x h q)                      ; occurs in the body
  (and (member x q) (not (member x h))))    ; but not in the head

(defun maj_locglob (h q)                    ; to classify the variables
  (mapc #'(lambda (x)
            (when (unsafe? x h q)
                  (setq *lvarloc (delete x *lvarloc))
                  (push x *lvarglob)))
        *lvarloc))
```

```
; mini-PrologII
; blocks.lsp
;

; I. Registers
;
(defconstant BottomFR 1)                    ; bottom of frozen goals stack
(defconstant BottomG 3000)                  ; bottom of global stack
(defconstant BottomL 6000)                  ; bottom of local stack
(defconstant BottomTR 10000)                ; bottom of trail
(defconstant A 12000)                       ; max of trail

(defvar Mem (make-array 12050 :initial-element 0))
(defvar FR)                                 ; top of frozen goals stack
(defvar TR)                                 ; top of trail
(defvar L)                                  ; top of local stack
(defvar G)                                  ; top of global stack
(defvar CP)                                 ; current continuation
```

```
(defvar CL)
(defvar Cut_pt)                          ; specific cut
(defvar FRCP)                            ; awakened goals
(defvar BL)                              ; backtracking registers
(defvar BG)
(defvar PC)                              ; currents goal and its
(defvar PCE)                             ; environments
(defvar PCG)
(defvar Duboulot)


(defmacro vset (v i x) '(setf (svref ,v ,i) ,x))

; II. Local Stack
;
;deterministic block [CL CP G Cut E]
;
(defmacro CL (b) '(svref Mem ,b))
(defmacro CP (b) '(svref Mem (1+ ,b)))
(defmacro G (b) '(svref Mem (+ ,b 2)))
(defmacro Cut (b) '(svref Mem (+ ,b 3)))
(defmacro E (b) '(+ ,b 4))

(defmacro push_cont ()                   ; saves continuation
  '(progn (vset Mem L CL) (vset Mem (1+ L) CP)))

(defmacro push_E (n)                     ; allocates local env
  '(let ((top (+ L 4 ,n)))
     (if (>= top BottomTR)
         (throw 'debord (print "Local Stack Overflow")))
     (vset Mem (+ L 3) Cut_pt)
     (dotimes (i ,n top)                 ; n local free variables
              (vset Mem (decf top) (cons 'LIBRE BottomG)))))

(defmacro maj_L (nl)                      ; updates top of local stack
  '(incf L (+ 4 ,nl)))


;choice-point : [a1 .. an A FR BCP BCL BG BL BP TR]
;
(defmacro TR (b) '(svref Mem (1- ,b)))
(defmacro BP (b) '(svref Mem (- ,b 2)))
(defmacro BL (b) '(svref Mem (- ,b 3)))
(defmacro BG (b) '(svref Mem (- ,b 4)))
(defmacro BCL (b) '(svref Mem (- ,b 5)))
(defmacro BCP (b) '(svref Mem (- ,b 6)))
(defmacro FR (b) '(svref Mem (- ,b 7)))
(defmacro A (b) '(svref Mem (- ,b 8)))

(defun save_args ()                       ; save args of current goal
  (dotimes (i (svref Mem A) (vset Mem (incf L i) i))
           (vset Mem (+ L i) (svref Mem (+ A i 1)))))
```

```
(defun push_choix ()                        ; allocates a choice point
  (save_args)                               ; goal args
  (vset Mem (incf L) FR)                    ; top of f.g. stack
  (vset Mem (incf L) CP)                    ; continuation
  (vset Mem (incf L) CL)
  (vset Mem (incf L) G)                     ; top of global stack
  (vset Mem (incf L) BL)                    ; last choice point
  (vset Mem (incf L 2) TR)                  ; top of trail
  (setq BL (incf L) BG G))                  ; new last choice point

(defun push_bpr (reste) (vset Mem (- BL 2) reste))

(defmacro size_C (b) '(+ 8 (A ,b)))

(defun pop_choix ()
  (setq L (- BL (size_C BL))                ; pops the block
        BL (BL BL)                          ; updates backtrack reg.
        BG (if (zerop BL) BottomG (BG BL))))

; III. Global Stack
;
(defmacro push_G (n)
  '(let ((top (+ G ,n)))
     (if (>= top BottomL)
         (throw 'debord (print "Global Stack Overflow")))
     (dotimes (i ,n (vset Mem (+ L 2) G)) ; n free global vars
              (vset Mem (decf top) (cons 'LIBRE BottomG)))))
(defmacro maj_G (n) '(incf G ,n))

;IV. Trail
;
(defmacro fgblock (x) '(cdr (svref Mem ,x)))

(defun pushtrail (x)
  (if (>= TR A) (throw 'debord (print "Trail Overflow")))
  (vset Mem TR x)
  (incf TR))

(defun poptrail (top)
  (do () ((= TR top))
      (let ((x (svref Mem (decf TR))))
        (if (numberp x)                     ; classical var else frozen
            (vset Mem x (cons 'LIBRE BottomG))
            (vset Mem (car x) (cons 'LIBRE (cdr x))))))))

; V. Frozen Goals Stack
;
(defmacro FGvar (x) '(svref Mem ,x))
(defmacro FGtail (x) '(svref Mem (1+ ,x)))
(defmacro FGgoal (x) '(svref Mem (+ 2 ,x)))
(defmacro FGenv (x) '(svref Mem (+ 3 ,x)))
(defmacro frozen? (x) '(< (fgblock ,x) BottomG))
```

```
(defmacro push_fg (v b eb r)              ; allocates a frozen block
  '(if (>= (+ FR 3) BottomG)
       (throw 'debord (print "Frozen Goals Stack Overflow"))
     (progn (vset Mem FR ,v)
            (vset Mem (incf FR) ,r)
            (vset Mem (incf FR) ,b)
            (vset Mem (incf FR) ,eb)
            (incf FR))))
```

```
; mini-PrologII
; unify.lsp
;
```

```
(defmacro bind (x sq e xt)                ; binds x to (sq,e)
  '(progn (if (or (and (> ,x BottomL) (< ,x BL)) (< ,x BG))
              (pushtrail ,xt))            ; if post-bound trail it
          (rplaca (svref Mem ,x) ,sq)
          (rplacd (svref Mem ,x) ,e)))
```

```
(defun bindte (x sq e)                    ; binds x to (sq,e)
  (if (frozen? x)
      (let ((y (fgblock x)))
        (push y FRCP)                     ; awakes the delayed goals
        (bind x sq e (cons x y)))         ; to trail the old value
    (bind x sq e x)))
```

```
(defun bindfg (x b eb r)                  ; binds x to the frozen goal b
  (bind x 'LIBRE FR (if (frozen? x) (cons x r) x))
  (push_fg x b eb r))                     ; allocates a new frozen block
```

```
(defun unify_with (largs el eg)           ; unifies head of clause
  (catch 'impossible                      ; with registers Ai
    (dotimes (i (svref Mem A))
             (unif
              (let ((te (svref Mem (+ A 1 i)))) (val (car te) (cdr te)))
              (ultimate (pop largs) el eg)))))
```

```lisp
; mini-PrologII
; resol.lsp
;

(defun forward ()
  (do () ((null Duboulot) (format t "no More~%"))
      (cond ((and (null CP)                 ; empty continuation
                  (null FRCP))              ; no awaken goals
             (answer))
            ((load_PC)                       ; selects first goal
             (cond
               ((user? PC)                   ; user defined
                (let ((d (def_of PC)))       ; associated set of clauses
                  (if d (pr2 d) (backtrack))))
               ((builtin? PC)                ; builtin
                (if (eq (apply (car PC) (cdr PC)) 'fail)
                    (backtrack)              ; evaluation fails
                    (cont_eval)))
               ((backtrack)))))))            ; undefined predicate

(defun load_A (largs el eg)                  ; loads Ai registers with
  (dotimes (i (length largs) (vset Mem A i)) ; goal's args
           (vset Mem (+ A i 1) (ultimate (pop largs) el eg))))

(defun load_PC ()
  (if FRCP
      (let ((x (())))
        (do ()                               ; sorts the goals depending on
            ((null FRCP))                    ; their freezing times
            (setq x (add_fg (pop FRCP) x)))
        (do ()                               ; allocates blocks in the
            ((null x))                       ; corresponding order
            (create_block (abs (pop x)))))
  (setq PC (pop CP) PCE (E CL) PCG (G CL) Cut_pt BL))

(defun other_fg (b r)
  (if (< (FGtail b) BottomG) (add_fg (FGtail b) r) r))

(defun add_fg (b r)                          ; sorts the various awakened
  (let ((b1 (if (numberp (FGgoal b)) (FGgoal b) b))) ; goals
    (if (eq (pred (FGgoal b1)) '|dif|)       ; dif first
        (insert (- b1) (other_fg b r))
        (let* ((v (svref Mem (FGvar b1)))     ; it is a freeze
               (te (val (car v) (cdr v))))
          (if (var? (car te))                 ; the var is still unbound
              (let ((y (adr (car te) (cdr te))))
                (bindfg y b1 nil (fgblock y)) ; delaying is perpetuated
                (other_fg b r))
              (insert b1 (other_fg b r))))))) ; insert the goal

(defun insert (b l)                          ; sorts the goals according to
  (if (or (null l) (> b (car l)))            ; their freezing times
```

```lisp
        (cons b l)
    (cons (car l) (insert b (cdr l)))))

(defmacro dec_goal (x)
  '(if (atom ,x) (list ,x) (cons (caar ,x) (cdr ,x))))

(defun create_block (b)                 ; block for an awakened goal
  (push_cont)                           ; saves current continuation
  (vset Mem (+ L 2) (FGenv b))          ; its global env
  (vset Mem (+ L 3) Cut_pt)             ; the corresponding cut point
  (setq CP (list (FGgoal b)) CL L)      ; new continuation
  (maj_L 0))                            ; ends the block

(defun pr2 (paq)                        ; proves PC
  (load_A (largs PC) PCE PCG)           ; loads its arguments
  (if CP
      (pr paq)
    (progn                              ; if last call and no interm.
      (if (<= BL CL) (setq L CL))       ; choice point then LCO
      (setq CP (CP CL) CL (CL CL))      ; next continuation
      (pr paq))))

(defun cont_eval ()
  (unless CP (if (<= BL CL) (setq L CL)) (setq CP (CP CL) CL (CL CL))))

(defun pr (paq)
  (if (cdr paq)                         ; alloc. choice point
      (progn (push_choix) (pr_choice paq))
    (pr_det (car paq))))                ; else deterministic

(defun pr_det (c)
  (if (eq (unify_with                   ; first tries to unify
           (largs (head c))
           (push_E (nloc c))
           (push_G (nglob c)))
          'fail)
      (backtrack)
    (progn                              ; success
      (maj_G (nglob c))                 ; terminates global env
      (when (tail c)                    ; c is rule
        (push_cont) saves current cont.
        (setq CP (tail c) CL L)         ; new one
        (maj_L (nloc c))))))            ; terminates local block

(defun pr_choice (paq)
  (let* ((resu (shallow_backtrack paq)) (c (car resu)) (r (cdr resu)))
    (cond ((null r)                     ; only one candidate remains
           (pop_choix)                  ; pops the rerun part
           (pr_det c))                  ; proof is now deterministic
          ( (push_bpr r)
            (maj_G (nglob c))
            (when (tail c)              ; c is a rule
```

```
              (push_cont)                    ; saves current cont.
              (setq CP (tail c) CL L) ; new one
              (maj_L (nloc c))))))) ; terminates local block

(defun shallow_backtrack (paq)          ; looks for a first success
  (if (and (cdr paq)                    ; more than one candidate
           (eq (unify_with             ; and unification fails
                (largs (head (car paq)))
                (push_E (nloc (car paq)))
                (push_G (nglob (car paq))))
               'fail))
      (progn
        (setq FRCP nil FR (FR BL))      ; restores registers
        (poptrail (TR BL))              ; restores env
        (shallow_backtrack (cdr paq)))
    paq))

(defun backtrack ()
  (if (zerop BL)                        ; no more choice points
      (setq Duboulot nil)               ; restores registers
    (progn (setq L BL G BG FR (FR L) FRCP nil Cut_pt (BL BL)
                 CP (BCP L) CL (BCL L) Cut_pt (BL BL))
           (load_A2)
           (poptrail (TR BL))           ; restores env
           (pr_choice (BP L)))))        ; restart the proof

(defun load_A2 ()                       ; restores Ai registers
  (let ((deb (- L (size_C L))))
    (dotimes (i (A L) (vset Mem A i))
             (vset Mem (+ A i 1) (svref Mem (+ deb i))))))

(defun myloop (c)
  (setq FR BottomFR G BottomG L BottomL TR BottomTR Cut_pt 0
        CP nil CL 0  BL 0 BG BottomG FRCP nil Duboulot t)
  (push_cont)                           ; initial continuation
  (push_E (nloc c))                     ; local env. for the query
  (push_G (nglob c))                    ; global env. for the query
  (setq CP (cdr c) CL L)                ; current continuation
  (maj_L (nloc c))                      ; ends local block
  (maj_G (nglob c)) (read-char)         ; ends global block
  (catch 'debord (forward))
  (myloop (read_prompt)))
```

```
; Mini-PrologII
; pred.lsp
;
(defvar Ob_Micro_Log
      '(|write| |nl| |tab| |read| |get| |get0|
        |var| |nonvar| |atomic| |atom| |number|
        ! |fail| |true|
        |divi| |mod| |plus| |minus| |mult| |le| |lt|
        |name| |consult| |abolish| |cputime| |statistics|
        |call| |freeze| |dif| |frozen_goals|))
(mapc #'(lambda (x) (setf (get x 'evaluable) t)) Ob_Micro_Log)

; !/0
(defun ! (n)
  (setq BL (Cut CL) BG (if (zerop BL) BottomG (BG BL))
        L (+ CL 4 n)))                        ; updates local stack

; call/1 (+term)
(defun |call| (x)
  (if (var? x)
      (let ((te (ultimate x PCE PCG)))  ; dereferences it
        (unless CP
                (if (<= BL CL) (setq L CL)) ; applies LCO
                (setq CP (CP CL) CL (CL CL))) ; new continuation
        (push_cont)                         ; saves continuation
        (vset Mem (+ L 2) (cdr te))         ; global env.
        (vset Mem (+ L 3) Cut_pt)           ; cut point
        (setq CP (list (dec_goal (car te))) CL L)
        (maj_L 0))                          ; ends local block
    (push (dec_goal x) CP)))                 ; adds it to CP

; freeze/2 (?var,+term)
(defun |freeze| (x p)
  (let ((xte (ultimate x PCE PCG)))         ; dereferences the var
    (if (var? (car xte))                    ; unbound
        (let ((y (adr (car xte) (cdr xte))) ; the location
              (pte (ultimate p PCE PCG)))   ; dereferences the goal
          (bindfg y (dec_goal (car pte)) (cdr pte) (fgblock y)))
      (|call| p))))                         ; else call p

; dif/2 (?term,?term)
(defun |dif| (x y)
  (let ((BL L) (BG G) (str TR) (FRCP nil)) ; saves registers
    (if (eq (uni x y) 'fail)               ; unification fails
        (poptrail str)                     ; restores env and succeeds
      (if (/= TR str)                      ; one var bound
          (let* ((xv (svref Mem (1- TR)))  ; selects one var
                 (v (if (numberp xv) xv (car xv))))
            (poptrail str)                 ; restores env
            (bindfg v PC PCG (fgblock v))) ; perpetuates the delaying
        'fail))))                          ; fails if equals
```

```
; statistics/0
(defun |statistics| ()
  (format t " local stack : ~A (~A used)~%"
          (- BottomTR BottomL) (- L BottomL))
  (format t " global stack : ~A (~A used)~%"
          (- BottomL BottomG) (- G BottomG))
  (format t " trail : ~A (~A used)~%"
          (- A BottomTR) (- TR BottomTR))
  (format t " frozen-goals stack : ~A (~A used)~%"
          BottomG (- FR BottomFR)))

; frozen_goals/0
(defun |frozen_goals| ()
  (do ((i (- FR 4) (- i 4)))            ; scans the frozen goals stack
      ((< i 0))
      (if (eq (car (svref Mem (FGvar i))) 'LIBRE) ; unbound
          (let ((b (if (numberp (FGgoal i)) (FGgoal i) i)))
            (writesf (pred (FGgoal b)) (largs (FGgoal b)) (FGenv b))
            (format t " frozen upon X~A~%" (FGvar i))))))




%
% Startup Mini-PrologII
%

$ eq(X,X).
$ neq(X,X):- !, fail.
$ neq(X,Y).

$ not(X) :- call(X), !, fail.
$ not(X).

$ repeat.
$ repeat :- repeat.

$ different(L) :- freeze(L,different1(L)).
$ different1([]).
$ different1([T|Q]) :- out_of(T,Q), different(Q).
$ out_of(X,L) :- freeze(L,out_of1(X,L)).
$ out_of1(X,[]).
$ out_of1(X,[T|Q]) :- dif(X,T), out_of(X,Q).

$ freeze2_or(X,Y,B) :- freeze(X,once(B,V)),
                       freeze(Y,once(B,V)).
$ once(B,V) :- var(V), !, call(B), eq(V,done).
$ once(B,V).
```

```
$ conc([],X,X) .
$ conc([T|Q],L,[T|R]) :- conc(Q,L,R).
$ member(X,[X|Y]).
$ member(X,[Y|Z]) :- member(X,Z).
$ del(X,[X|Y],Y).
$ del(X,[Y|Z],[Y|R]) :- del(X,Z,R).

$ wb(X) :- write(X), write(' ').
$ wf(X) :- write(X), nl, fail.
```

APPENDIX D

COMMON PART

```
; mini-Cprolog & mini-PrologII
; cparser.lsp
;

(defvar *lvarloc nil)                    ; list of local vars
(defvar *lvarglob nil)                   ; list of global vars
(set-macro-character #\% (get-macro-character #\;))

(defun rch ()                            ; skips Newline
  (do ((ch (read-char) (read-char)))
      ((char/= ch #\Newline) ch)))
(defun rchnsep ()                        ; skips Newline and Space
  (do ((ch (rch) (rch)))
      ((char/= ch #\space) ch)))

(defun special (ch) (char= ch '#\_))
(defun alphanum (ch)                     ; symbol normal constituant
  (or (alphanumericp ch) (special ch)))
(defun valdigit (ch) (digit-char-p ch))

(defun read_number (ch)                  ; next integer
  (do ((v (valdigit ch) (+ (* v 10) (valdigit (read-char)))))
      ((not (digit-char-p (peek-char))) v)))

(defun implode (lch) (intern (map 'string #'identity lch)))

(defun read_atom (ch)                    ; next normal symbol
  (do ((lch (list ch) (push (read-char) lch)))
      ((not (alphanum (peek-char))) (implode (reverse lch)))))

(defun read_at (ch)                      ; next special symbol
  (do ((lch (list ch) (push (read-char) lch)))
      ((char= (peek-char) #\') (read-char) (implode (reverse lch)))))

(defun read_string (ch)                  ; Prolog list of ascii codes
  (do ((lch (list (char-int ch)) (push (char-int (read-char)) lch)))
      ((char= (peek-char) #\") (read-char) (do_l (reverse lch)))))

(defun read_var (ch n)                   ; next variable
  (status (read_atom ch) n))
```

```
(defun status (nom n)                   ; n is the term's depth
  (if (= n 1)                           ; local if not yet global
      (unless (member nom *lvarglob) (pushnew nom *lvarloc))
      (progn (if (member nom *lvarloc)  ; else global
                 (setq *lvarloc (delete nom *lvarloc)))
             (pushnew nom *lvarglob)))
  nom)

(defun read_simple (ch n)               ; next simple term
  (cond
    ((or (upper-case-p ch) (special ch)) (read_var ch n))
    ((digit-char-p ch) (read_number ch))
    ((char= ch #\") (read_string (read-char)))
    ((char= ch #\') (read_at (read-char)))
    (t (read_atom ch))))

(defun read_fct (ch n)                  ; next functional term
  (let ((fct (read_simple ch n)) (c (rchnsep)))
    (if (char= c #\()                   ; reads its arguments
        (let ((la (read_args (rchnsep) (1+ n))))
          (cons (list fct (length la)) la)) ; adds its descriptor
      (progn (unread-char c) fct))))    ; else atom

(defun read_args (ch n)                 ; args of a functional term
  (let ((arg (read_term ch n)))
    (if (char= (rchnsep) #\,)
        (cons arg (read_args (rchnsep) n))
      (list arg))))

(defun read_list (ch n)                 ; next list
  (if (char= ch #\])                    ; ending with the empty list
      ()
    (let ((te (read_term ch n)))        ; gets the head
      (case (rchnsep)
            (#\, (list '(\. 2) te (read_list (rchnsep) n)))
            (#\| (prog1                 ; dotted pair
                    (list '(\. 2) te (read_term (rchnsep) n))
                  (rchnsep)))
            (#\] (list '(\. 2) te nil))))))

(defun read_term (ch n)                 ; next term
  (if (char= ch #\[) (read_list (rchnsep) (1+ n)) (read_fct ch n)))

(defun read_tail (ch)                   ; read the body of a rule
  (let ((tete (read_pred ch)))          ; gets the first goal
    (if (char= (rchnsep) #\.)
        (list tete)
      (cons tete (read_tail (rchnsep))))))

(defun read_clause (ch)                 ; reads a clause
  (let ((tete (read_pred ch)))
    (if (char= (rchnsep) #\.)
```

```
        (list tete)
      (progn (read-char) (cons tete (read_tail (rchnsep)))))))))

(defun c (l)                                  ; codes once read
  (if (atom l)
      (if (member l *lvarloc)
          (cons 'L (position l *lvarloc))
          (if (member l *lvarglob) (cons 'G (position l *lvarglob)) l))
    (if (eq (car l) '!)                       ; the cut with its intern
        (list '! (length *lvarloc))           ; argument
      (cons (c (car l)) (c (cdr l))))))
```

```
; miniCprolog & Mini-PrologII
; cutili.lsp
;

(defmacro nloc (c) '(caar ,c))            ; number of local vars
(defmacro nglob (c) '(cdar ,c))           ; number of global vars
(defmacro head (c) '(cadr ,c))            ; head of a clause
(defmacro tail (c) '(cddr ,c))            ; body of a clause
(defmacro pred (g) '(car ,g))             ; predicate symbol
(defmacro largs (g) '(cdr ,g))            ; list of arguments

(defmacro functor (des) '(car ,des))
(defmacro arity (des) '(cadr ,des))
(defmacro des (te) '(car ,te))            ; functor/arity
(defmacro var? (v) '(and (consp ,v) (numberp (cdr ,v))))
(defmacro list? (x) '(eq (functor (des ,x)) '\.))

(defmacro user? (g)                       ; user defined predicate
  '(get (pred ,g) 'def))
(defmacro builtin? (g)                    ; builtin predicate
  '(get (pred ,g) 'evaluable))
(defmacro def_of (g)                      ; gets the subset of clauses
  '(get (pred ,g)                         ; depending on the nature of
        (if (largs ,g)                    ; the first arg of goal PC
            (nature (car (ultimate (car (largs ,g)) PCE PCG)))
          'def)))

(defun nature (te)                        ; determines the associated
  (cond                                   ; subset according to
   ((var? te) 'def)                       ; clause indexing
   ((null te) 'empty)
   ((atom te) 'atom)
   ((list? te) 'list)
   (t 'fonct)))
```

```
(defun add_cl (pred c ind)              ; adds a clause to the end
  (setf (get pred ind) (append (get pred ind) (list c))))

(set-macro-character                    ; to automatically read,
 #\$                                    ; code and index
 #'(lambda (stream char)
     (let* ( (*standard-input* stream) (c (read_code_cl)))
       (add_cl (pred (head c)) c 'def)  ; always in the var subset
       (if (largs (head c))
           (let ((b (nature (car (largs (head c))))))
             (if (eq b 'def)            ; first arg is a variable
                 (mapc                  ; add c to all the subsets
                  #' (lambda (x) (add_cl (pred (head c)) c x))
                  '(atom empty list fonct))
                 (add_cl (pred (head c)) c b)))) ; add to only one subset
       (values)))

(defun answer ()                        ; prints the values of vars
  (printvar)                            ; occurring in the query
  (if (zerop BL)                        ; no more choice point
      (setq Duboulot nil)
    (if (and (princ "More : ")          ; asks the user
             (string= (read-line) ";")) ; if necessary
        (backtrack)                     ; forces backtracking
      (setq Duboulot nil))))

(defun printvar ()
  (if (and (null *lvarloc)              ; ground query ?
           (null *lvarglob))
      (format t "Yes ~%")               ; success
    (let ((nl -1) (ng -1))
      (mapc                             ; first, local variables
       #' (lambda (x)
            (format t "~A = " x)
            (write1 (ult (cons 'L (incf nl)) (E BottomL))) (terpri))
       *lvarloc)
      (mapc                             ; then global variables
       #' (lambda (x)
            (format t "~A = " x)
            (write1 (ult (cons 'G (incf ng)) BottomG)) (terpri))
       *lvarglob))))
```

```
; mini-Cprolog & mini-PrologII
; cunify.lsp
;
(defmacro adr (v e) '(+ (cdr ,v) ,e))
(defmacro value (v e) '(svref Mem (adr ,v ,e)))

(defun ult (v e)                        ; dereferences variable v
  (let ((te (value v e)))               ; in environment e
    (cond
      ((eq (car te) 'LIBRE) (cons v e)) ; if unbound, itself
      ((var? (car te)) (ult (car te) (cdr te)))
      ( te))))                          ; its value

(defun val (x e)                        ; generalises to a term
  (if (var? x) (ult x e) (cons x e)))

(defun ultimate (x el eg)               ; idem but selects env
  (if (var? x)
      (if (eq (car x) 'L) (ult x el) (ult x eg))
    (cons x eg)))

(defmacro bindv (x ex y ey)             ; L2 Binding
  '(let ((ax (adr ,x ,ex)) (ay (adr ,y ,ey)))
     (if (< ax ay)                      ; the younger one is always
         (bindte ay ,x ,ex)             ; bound to the senior one
       (bindte ax ,y ,ey))))

(defun unif (t1 t2)                     ; unify two terms t1 and t2
  (let ((x (car t1)) (ex (cdr t1)) (y (car t2)) (ey (cdr t2)))
    (cond
      ((var? y)
       (if (var? x)                     ; two variables
           (if (= (adr x ex) (adr y ey)) t (bindv y ey x ex))
         (bindte (adr y ey) x ex)))     ; binds y
      ((var? x) (bindte (adr x ex) y ey))
      ((and (atom x) (atom y))          ; two constants
       (if (eql x y) t (throw 'impossible 'fail)))
      ((or (atom x) (atom y)) (throw 'impossible 'fail))
      ( (let ((dx (pop x)) (dy (pop y))) ; two structured terms
          (if (and (eq (functor dx) (functor dy))
                   (= (arity dx) (arity dy)))
              (do () ((null x))          ; same functor and arity
                  (unif (val (pop x) ex) (val (pop y) ey)))
            (throw 'impossible 'fail)))))))
```

```
; mini-Cprolog & mini-PrologII
; cpred.lsp
;

(defmacro value1 (x) '(car (ultimate ,x PCE PCG)))
(defun uni (x y)
  (catch 'impossible
    (unif (ultimate x PCE PCG) (ultimate y PCE PCG))))
                                          ;write/1 (?term)
(defun |write| (x)
  (write1 (ultimate x PCE PCG)))
(defun write1 (te)                        ; depending on te
  (let ((x (car te)) (e (cdr te)))
    (cond
      ((null x) (format t "[]"))
      ((atom x) (format t "~A" x))
      ((var? x) (format t "X~A" (adr x e)))
      ((list? x) (format t "["
        (writesl (val (cadr x) e) (val (caddr x) e))
        (format t "]"))
      ((writesf (functor (des x)) (largs x) e)))))
(defun writesl (te r)                     ; for a list
  (write1 te)
  (let ((q (car r)) (e (cdr r)))
    (cond
      ((null q))
      ((var? q) (format t "|X~A" (adr q e)))
      (t (format t ",")
        (writesl (val (cadr q) e) (val (caddr q) e)))))))
(defun writesf (fct largs e)              ; for a functional term
  (format t "~A(" fct)
  (write1 (val (car largs) e))
  (mapc #' (lambda (x) (format t ",") (write1 (val x e)))
        (cdr largs))
  (format t ")"))
                                    ;nl/0
(defun |nl| () (terpri))
                                    ;tab/1 (+int)
(defun |tab| (x)
  (dotimes (i (value1 x)) (format t " ")))
                                    ;read/1 (?term)
(defun |read| (x)
  (let ((te (read_terme)))          ; gets the term
    (catch 'impossible
      (unif (ultimate x PCE PCG)
            (cons (cdr te) (push1_g (car te)))))))
(defun read_terme ()
  (let ((*lvarloc nil) (*lvarglob nil))
    (let ((te (read_term (rchnsep) 2)))
      (rchnsep) (cons (length *lvarglob) (c te)))))
```

```
(defun push1_g (n)
  (if (>= (+ G n) BottomL)                   ; allocates a global env
      (throw 'debord (print "Global Stack Overflow")))
  (dotimes (i n (- G n))
          (vset Mem G (cons 'LIBRE BottomG))
          (incf G)))
                                             ;get/1 (?car)
(defun |get| (x)
  (uni x (char-int (rchnsep))))
                                             ;get0/1 (?car)
(defun |get0| (x)
  (uni x (char-int (read-char))))
                                             ;var/1 (?term)
(defun |var| (x)
  (unless (var? (value1 x)) 'fail))
                                             ;nonvar/1 (?term)
(defun |nonvar| (x)
  (if (var? (value1 x)) 'fail))
                                             ;atomic/1 (?term)
(defun |atomic| (x)
  (if (listp (value1 x)) 'fail))
                                             ;atom/1 (?term)
(defun |atom| (x)
  (unless (symbolp (value1 x)) 'fail))
                                             ;number/1 (?term)
(defun |number| (x)
  (unless (numberp (value1 x)) 'fail))
                                             ;fail/0
(defun |fail| () 'fail)
                                             ;true/0
(defun |true| ())
                                             ;divi/3 (+int,+int,?int)
(defun |divi| (x y z)
  (uni z (floor (value1 x) (value1 y))))
                                             ;mod/3 (+int,+int,?int)
(defun |mod| (x y z)
  (uni z (rem (value1 x) (value1 y))))
                                             ;plus/3 (+int,+int,?int)
(defun |plus| (x y z)
  (uni z (+ (value1 x) (value1 y))))
                                             ;minus/3 (+int,+int,?int)
(defun |minus| (x y z)
  (uni z (- (value1 x) (value1 y))))
                                             ;mult/3 (+int,+int,?int)
(defun |mult| (x y z)
  (uni z (* (value1 x) (value1 y))))
                                             ;le/2 (+int,+int)
(defun |le| (x y)
  (if (> (value1 x) (value1 y)) 'fail))
```

```
(defun |lt| (x y)
  (if (>= (value1 x) (value1 y)) 'fail))
                                      ;name/2 (?atom,?list)
(defun |name| (x y)
  (let ((b (value1 x)))
    (if (var? b)
        (uni x (impl (undo_1 (ultimate y PCE PCG))))
        (uni y (do_1 (expl b))))))

(defun undo_1 (te)
  (let ((x (car te)) (e (cdr te)))
    (if (atom x)
        x
      (cons (undo_1 (val (cadr x) e)) (undo_1 (val (caddr x) e))))))
(defun do_1 (x)
  (if (atom x) x (list '(\. 2) (car x) (do_1 (cdr x)))))
(defun impl (l)
  (intern (map 'string #'int-char l)))
(defun expl (at)
  (map 'list #'char-int (string at)))


                                      ;consult/1 (+atom)
(defun |consult| (f)
  (format t "~A~%" (load (value1 f))))
                                      ; abolish/1 (+ atom)
(defun |abolish| (p)
  (mapc #'(lambda (x) (setf (get p x) nil))
        '(atom empty list fonct def)))
                                      ; cputime/1 (? int)
(defun |cputime| (x)
  (uni x (float (/ (get-internal-run-time)
                   internal-time-units-per-second))))
```

BIBLIOGRAPHY

[AAI88]   *AAIS-Prolog M2.0*. Mountain View, Calif.: AAIS Inc., 1988.

[ACHS88]   K. Appleby, M. Carlsson, S. Haridi, and D. Sahlin. Garbage-Collection for Prolog Based on WAM. *CACM* 31, no. 6 (1988): 719–741.

[AE82]   K. R. Apt and M. H. Van Emden. Contribution to the Theory of Logic Programming. *JACM* 29, no. 3 (1982): 841–862.

[ASS85]   H. Abelson, G. Sussman, and J. Sussman. *Structure and Interpretation of Computer Programs*. MIT Press, 1985.

[Bak78]   H. Baker. List Processing in Real Time on a Computer. *CACM* 21, no. 4 (1978): 280–294.

[Bal83]   G. Ballieu. A Virtual Machine to Implement Prolog. In *Logic Programming Workshop '83*, pp. 40–52. Algarve, Portugal, 1983.

[BBC83]   D. L. Bowen, L. M. Byrd, and W. F. Clocksin. A Portable Prolog Compiler. In *Logic Programming Workshop '83*, pp. 74–83.

[BBCT86]   K. Bowen, A. Buettner, I. Cicekli, and A. Turk. The Design and Implementation of a High-Speed Incrementable Portable Prolog Compiler. In E. Shapiro, ed. *Third International Logic Programming Conference*, pp. 650–656, LNCS 225. Springer-Verlag, 1986.

[BC83]   M. Bidoit and J. Corbin. Pour la réhabilitation de l'algorithme d'unification de Robinson. C. R. Académie des Sciences de Paris, February 1983.

[BCRU84]   Y. Bekkers, B. Canet, O. Ridoux, and L. Ungaro. A Memory Management Machine for Prolog Interpreters. In S. A. Tarnlund, ed. *Second International Logic Programming Conference*, pp. 343–351. Uppsala, Sweden, 1984.

[BCRU85]   ———. *Le projet MALI: Bilan et Perspectives*. Technical Report. IRISA. Rennes, 1985.

[BCRU86]   ———. MALI a Memory with a Real Time Garbage-Collector for Implementing Logic Programming Languages. In *Third IEEE Symposium on Logic Programming*. Salt Lake City, 1986.

[BCU82]   Y. Bekkers, B. Canet, and L. Ungaro. Problèmes de gestion mémoire dans les interprètes Prolog. In M. Dincbas and D. Feuerstein, eds. *Journées Programmation en Logique du CNET*, pp. 20–26. 1982.

[BDL84]   S. Bourgault, M. Dincbas, and J. P. Lepape. *Manuel Lislog*. Lannion, France: CNET, 1984.

[Bes85]   P. Besnard. Sur la détection de boucles infinies en Programmation en Logique. In S. Bourgault and M. Dincbas, eds. *Journées Programmation en Logique du CNET*, pp. 233–246. 1985.

[Bil84]   M. Billaud. *Manuel d'utilisation de Prolog-Bordeaux*. Technical Report. Greco de Programmation. Université de Bordeaux. France, 1984.

[Bil85]   ————. *Une formalisation des Structures de controle de Prolog*. Ph.D. thesis, Université de Bordeaux, 1985.

[Bil86]   ————. *Prolog Control Structures: A Formalisation and Its Application*. Technical Report 8627. Université de Bordeaux, 1986.

[BIM86]   *BIM-Prolog manual*. Everberg, Belgium: BIM, 1986.

[BJM83]   G. Barberye, T. Joubert, and M. Martin. *Manuel d'utilisation du Prolog Cnet*. Paris: CNET, note nt/paa/clc/lsc/1058 ed., 1983.

[BK82]   K. A. Bowen and R. A. Kowalski. Amalgaming Language and Meta-Language in Logic Programming. In K. L. Clark and S. A. Tarnlund, eds. *Logic Programming*, pp. 153–172. Academic Press, 1982.

[BM72]   R. S. Boyer and J. S. Moore. The Sharing of Structures in Theorem Proving Programs. In *Machine Intelligence 7*, pp. 101–116. Edinburgh University Press, 1972.

[BM73]   G. Battani and H. Meloni. *Interpréteur du langage de programmation Prolog*. Technical Report, G.I.A. Université Aix-Marseille, 1973.

[Boi84]   P. Boizumault. Un modèle de trace pour Prolog. In M. Dincbas, ed. *Journées Programmation en Logique du CNET*, pages 61–72. 1984.

[Boi85]   ————. *Etude de l'interprétation de Prolog, Réalisations en Lisp*. Ph.D. thesis, Université Paris VI, 1985.

[Boi86a]   ————. A Classical Implementation for Prolog-II. In *ESOP86*, pp. 262–273. LNCS 213. Springer-Verlag, 1986.

[Boi86b]   P. Boizumault. A General Model to Implement Dif and Freeze. In E. Shapiro, ed. *Third International Logic Programming Conference*, pp. 585–592. LNCS 225. Springer-Verlag, 1986.

[BP81]   M. Bruynooghe and L. M. Pereira. *Revision of Top-Down Logical Reasoning through Intelligent Backtracking*. Technical Report cw23. Katholieke Universiteit Leuven, 1981.

[BP84]   ————. Deduction Revision by Intelligent Backtracking. In J. A. Campbell, ed. *Implementations of Prolog*, pp. 194–215. Hellis Horwood, 1984.

[Bra86]   I. Bratko. *Prolog Programming for Artificial Intelligence*. Addison Wesley, 1986.

[Bru76]   M. Bruynooghe. *An Interpreter for Predicate Logic Programs*. Technical Report Cw10. Katholieke Universiteit Leuven, 1976.

[Bru80]   ————. The Memory Management of Prolog Implementations. In *Logic Programming Workshop '80*, pp. 12–20. Debrecen, Hungary, 1980.

[Bru81]   ———. Intelligent Backtracking for an Interpreter of Horn Clauses Logic Programs. In B. Domolki and T. Gergely, eds. *Mathematical Logic in Computer Science*, pp. 215–258. North Holland, 1981.

[Bru82a]   ———. The Memory Management of Prolog Implementation. In K. L. Clark and S. A. Tarnlund, eds. *Logic Programming*, pp. 83–98. Academic Press, 1982.

[Bru82b]   ———. A Note on Garbage-Collection in Prolog Interpreters. In M. Van Caneghem, ed. *First International Logic Programming Conference*, pp. 52–55. 1982.

[Bue86]   K. Buettner. Fast Decompilation of Compiled Prolog Clauses. In E. Shapiro, ed. *Third International Logic Programming Conference*, pp. 663–670. LNCS 225. Springer-Verlag, 1986.

[Byr79]   L. Byrd. *The Prolog-DEC10 Debugging Package*. Technical Report. D.A.I. University of Edinburgh, 1979.

[Byr80]   ———. Understanding the Control Flow of Prolog Programs. In *Logic Programming Workshop '80*, pp. 127–138. Debrecen, Hungary, 1980.

[Cab81]   F. G. Mac Cabe. *Micro-Prolog Programmer's Reference Manual*. Logic Programming Associates, 1981.

[Can82]   M. Van Caneghem. *Manuel d'utilisation de Prolog-II*. Technical Report. G.I.A. Faculté des Sciences, Université Aix-Marseille, 1982.

[Can84]   ———. *L'anatomie de Prolog*. Ph.D. thesis, Faculté des Sciences de Luminy. Marseille, 1984.

[Can86]   ———. *L'anatomie de Prolog*. InterEditions, 1986.

[Car86]   M. Carlsson. *Compilation for Tricia and its Abstract Machine*. Technical Report 35. Uppsala University, Sweden, 1986.

[Car87]   ———. Freeze, Indexing and Other Implementation Issues in the WAM. In J. L. Lassez, ed. *Fourth International Logic Programming Conference*, pp. 40–58. MIT Press, 1987.

[CC79]   K. L. Clark and F. G. Mac Cabe. The Control Facilities of IC-Prolog. In D. Mitchie, ed. *Expert Systems in the Microelectronic Age*, pp. 153–167. Edinburgh University Press, 1979.

[CC80]   ———. IC-Prolog: Aspects of Its Implementation. In *Logic Programming Workshop '80*, pp. 190–197. Debrecen, Hungary, 1980.

[CC84]   ———. *Micro-Prolog : Programming in Logic*. Prentice-Hall, 1984.

[CCF86]   C. Codognet, P. Codognet, and G. File. A Very Intelligent Backtracking Method for Logic Programs. In *European Symposium on Programming*, pp. 315–326. LNCS 213. Springer-Verlag, 1986.

[CCF88]   ———. Yet Another Intelligent Backtracking Method. In R. A. Kowalski and K. A. Bowen, eds. *Fifth International Logic Programming Conference*, pp. 447–465. MIT Press, 1988.

[CCG82]   K. L. Clark, F. G. MacCabe, and S. Gregory. IC-Prolog Language Features. In K. L. Clark and S. A. Tarnlund, eds. *Logic Programming*, pp. 253–266. Academic Press, 1982.

[CCP80]   H. Coelho, J. C. Cotta, and L. M. Pereira. *How to Solve It with Prolog.* Laboratorio Nacional de Engenharia Civil. Lisbon, 1980.

[CG85]   K. Clark and S. Gregory. Note on the Implementation of Parlog. *Journal of Logic Programming* 2, no. 1 (1985): 17–42.

[CH87]   J. Cohen and T. Hickey. Parsing and Compiling Using Prolog. *ACM Transactions on Programming Languages and Systems* 9, no. 2 (1987): 125–163.

[CHR87]   L. Chevallier, S. Le Huitouze, and O. Ridoux. Style de programmation pour une machine de programmation logique munie d'un récupérateur de mémoire. In S. Bourgault and M. Dincbas, eds. *Journées Programmation en Logique du CNET*, pp. 245–263. 1987.

[CK83]   M. Carlsson and K. Kahn. *LM-Prolog User Manual.* Technical Report 24. Uppsala University, Sweden, 1983.

[CKC79]   A. Colmerauer, H. Kanoui, and M. Van Caneghem. *Etude et réalisation d'un système Prolog.* Technical Report. Université d'Aix-Marseille, 1979.

[CKC83]   ———. Prolog, bases théoriques et développements actuels. *TSI* 2, no. 4 (1983): 271–311.

[CKPR72] ———, R. Pasero, and P. Roussel. *Un système de communication homme machine en francais.* Technical Report. G.I.A. Université d'Aix-Marseille, 1972.

[CL73]   C. L. Chang and R. C. Lee. *Symbolic Logic and Mechanical Theorem Proving.* Academic Press, 1973.

[Clo85]   W. F. Clocksin. Design and Simulation of a Sequential Prolog Machine. *New Generation Computing* 3, no. 1 (1985):101–120.

[CM84]   ———, and C. S. Mellish. *Programming in Prolog.* 2d ed. Springer-Verlag, 1984.

[Col82a] A. Colmerauer. Prolog and Infinite Trees. In K. L. Clark and S. A. Tarnlund, eds. *Logic Programming*, pp. 231–251. Academic Press, 1982.

[Col82b] ———. *Prolog-II, manuel de référence et modèle théorique.* Technical Report. G.I.A.. Faculté des Sciences, Université Aix-Marseille, 1982.

[Col84]   ———. Prolog langage de l'Intelligence Artificielle. *La Recherche* (September 1984): 1104–1114.

[Col87]   ———. Opening the PrologIII Universe. *Byte* (August 1987).

[Con86]   M. Condillac. *Prolog: fondements et applications.* Dunod, 1986.

[Cox84]   P. T. Cox. Finding Backtrack Points for Intelligent Backtracking. In J. A. Campbell, ed. *Implementations of Prolog*, pp. 216–233. Hellis Horwood, 1984.

[CP81]   ——— and T. Pietrzykowski. Deduction Plans: a Basis for Intelligent Backtracking. *IEEE Transactions on Pattern Analysis and Machine Intelligence* (1981): 52–65.

[DDP84]   T. P. Dobry, A. M. Despain, and Y. N. Patt. Design Decisions Influencing the Micro-Architecture for a Prolog Machine. In *17th Annual Micro-Programming Workshop*, pp. 217–231. 1984.

[Deb88]   S. K. Debray. *The SB-Prolog system, version 2.5.* Department of Computer Science, University of Arizona, 1988.

[Del86]   J. P. Delahaye. *Outils logiques pour l'Intelligence Artificielle.* Eyrolles, 1986.

[Din79]   M. Dincbas. *Le système de résolution de problèmes MetaLog.* Technical Report 3146, CERT-DERI. 1979.

[Din80]   ———. The MetaLog Problem Solving System. In *Logic Programming Workshop '80,* pp. 80–91. Debrecen, Hungary, 1980.

[Din85]   ———. Les langages combinant Lisp et Prolog. *Génie Logiciel* 2 (1985): 25–28.

[Diz89]   P. Saint Dizier. *An Introduction to Programming in Prolog.* Springer-Verlag, 1989.

[dKCRS89] J. Chassin de Kergommeaux, P. Codognet, P. Robert, and J. C. Syre. Une revue des modèles de programmation en logique parallèle. *TSI* 3 and 4 (1989).

[Don83]   P. Donz. *Manuel de référence Prolog-Criss.* Grenoble: Criss, 1983.

[DRM87]   P. Deransart, G. Richard, and C. Moss. Spécification formelle de Prolog standard. In S. Bourgault and M. Dincbas, eds. *Journées Programmation en Logique du CNET,* pp. 455–482. 1987.

[DSH87]   M. Dincbas, H. Simonis, and P. Van Hentenryck. Extending Equation Solving and Constraint Handling in Logic Programming. In *Colloquim on the Resolution of Equations in Algebraic Structures.* 1987.

[Duc86]   M. Ducasse. Opium: un outil de trace sophistiqué pour Prolog. In S. Bourgault and M. Dincbas, eds. *Journées Programmation en Logique du CNET,* pp. 281–292. 1986.

[Eis84]   M. Eisenstadt. A Powerful Prolog Trace Package. In T. O'Shea, ed. *ECAI84,* pp. 515–524. North Holland, 1984.

[Elc83]   W. E. Elcock. The Pragmatics of Prolog: Some Comments. In *Logic Programming Workshop,* pp. 94–106. Algarve, Portugal, 1983.

[Emd84]   M. H. Van Emden. An Interpreting Algorithm for Prolog Programs. In J. A. Campbell, ed. *Implementations of Prolog,* pp. 93–110. Ellis Horwood, 1984.

[Fer86]   J. Ferber. Prolog: l'interprétation par effacement de buts. *Micro-Systèmes* (September 1986): 190–195.

[FF86]   K. Fuchi and K. Furukawa. The Role of Logic Programming in the Fifth-Generation Computer Project. In E. Shapiro, ed. *Third International Logic Programming Conference,* pp. 1–24. LNCS 225. Springer-Verlag, 1986.

[FGP*85]   N. Francez, S. Goldenberg, R. Pinter, M. Tiomkin, and S. Tur. An environment for Logic Programming. *JACM* 165, no. 2 (1985):179–190.

[Gal85]   H. Gallaire. Logic Programming: Further Developments. In *2nd IEEE Symposium on Logic Programming.* Boston, 1985.

[GJ89]   A. Marien G. Janssens, B. Demoen. Improving Register Allocation of WAM by Reordering Unification. In R. Kowalski and K. A. Bowen, eds. *Fifth International Logic Programming Conference*, pp. 1388–1402. MIT Press, 1989.

[GKPC86]   F. Giannesini, H. Kanoui, R. Pasero, and M. Van Caneghem. *Prolog*. Addison Wesley, 1986.

[GL82]   H. Gallaire and C. Lasserre. Metalevel Control for Logic Programs. In K. L. Clark and S. A. Tarnlund, eds. *Logic Programming*, pp. 173–185. Academic Press, 1982.

[GLD89]   A. Gal, G. Lapalme, and P. Saint Dizier. *Prolog pour l'analyse automatique du Langage Naturel*. Eyrolles, 1989.

[Glo84]   P. Y. Gloess. Logis : Un système Prolog dans un environnement Lisp. In M. Dincbas, ed. *Journées Programmation en Logique du CNET*, pp. 213–222. 1984.

[GMP87]   J. Gee, S. Melvin, and Y. N. Patt. Advantages of Implementing Prolog by Microprogramming a Host General Purpose Computer. In J. L. Lassez, ed. *Fourth International Logic Programming Conference*, pp. 1–20. MIT Press, 1987.

[GN87]   M. R. Genesereth and N. J. Nilsson. *Logical Foundations of Artificial Intelligence*. Morgan Kaufmann, 1987.

[Gre77]   P. Greussay. *Contribution à la définition interprétative et à l'implémentation des lambda-langages*. Ph.D. thesis, Université Paris VI, 1977.

[Gre82]   ———. *Le système Vlisp*. 1982.

[Gre83]   ———. Un mariage heureux entre Lisp et Prolog. In M. Dincbas, ed. *Journées Programmation en Logique du CNET*, pp. 111–121. 1983.

[Gre85]   ———. X-Prolog-II, un Prolog-II expérimental. Documentation on-line. LITP. Paris, 1985.

[Hen89]   P. Van Hentenryck. *Constraint Satisfaction in Logic Programming*. MIT Press, 1989.

[Her30]   J. Herbrand. *Recherches sur la théorie de la démonstration*. Ph.D. thesis, Paris, 1930.

[Hog84]   C. J. Hogger. *Introduction to Logic Programming*. Apic Studies in Data Processing. Academic Press, 1984.

[JL87]   J. Jaffar and J. L. Lassez. Constraint Logic Programmming. In *14th ACM POPL Symposium*. 1987.

[Kan82]   H. Kanoui. *Manuel d'exemples de Prolog-II*. Technical Report. G.I.A. Faculté des Sciences, Université Aix-Marseille, 1982.

[KC84]   K. Kahn and M. Carlsson. How to Implement Prolog on a Lisp Machine. In J. A. Campbell, ed. *Implementations of Prolog*, pp. 117–134. Ellis Horwood, 1984.

[Kom82]   J. Komorowski. Qlog, the Programming Environment for Prolog in Lisp. In K. L. Clark and S. A. Tarnlund, eds. *Logic Programming*, pp. 315–322. Academic Press, 1982.

[Kow74]   R. A. Kowalski. Predicate Logic as a Programming Language. In *Proceedings of IFIP 74*, pp. 569–574. North Holland, 1974.

[Kow79]   ———. Algorithm = Logic + Control. *Communications of the ACM 22*, no. 7 (1979): 424–436.

[Kow80]   ———. *Logic for Problem Solving*. North Holland, 1980.

[KS85]   F. Kluzniak and S. Szpakowics. *Prolog for Programmers. Apic Studies in Data Processing 24*. Academic Press, 1985.

[Lep85]   B. Lepape. Panorama des mises en oeuvre du langage Prolog. *Revue Génie Logiciel 2* (1985):10–17.

[LG80]   C. Lasserre and H. Gallaire. Controlling Backtrack in Horn Clauses Programming. In *Logic Programming Workshop*, pp. 286–292. Debrecen, Hungary, 1980.

[LL86]   P. Lebaube and B. Lepape. Prolog et les techniques de controle. In S. Bourgault and M. Dincbas, eds. *Journées Programmation en Logique du CNET*, pp. 51–74. 1986.

[Llo86]   J. W. Lloyd. *Foundations of Logic Programming*. 2d ed. Springer-Verlag, 1986.

[Loi87]   V. Loia. *Clog: Yet Another Prolog Interpreter*. Technical Report. LITP. Université Paris VI, 1987.

[Lov78]   D. W. Loveland. *Automated Theorem Proving: A Logical Basis*. North Holland, 1978.

[Mal87]   J. Malpas. *Prolog: A Relational Language and Its Applications*. Prentice Hall, 1987.

[Man84]   *Manuel de référence, Prolog/P v2-10*. Puteaux, France: Cril, 1984.

[Man85]   *Manuel de référence Prolog-Criss*. Alma, Saint Martin d'Hères: Criss, 1985.

[Man86]   *Manuel de référence Turbo-Prolog*. Sèvres, France: Borland, 1986.

[Man87]   *Manuel de référence Delphia-Prolog*. Grenoble, France: Delphia, 1987.

[Mar88]   A. Marien. An Optimal Intermediate Code for Structure Creation in a WAM-Based Prolog Implementation. In *International Computer Science Conference*, pp. 229–236. 1988.

[Mel79]   C. S. Mellish. *Short Guide to Unix Prolog Implementation*. Technical Report. D.A.I. Edinburgh, 1979.

[Mel80]   ———. An Alternative to Structure-Sharing in the Implementation of a Prolog Interpreter. In *Logic Programming Workshop*, pp. 21–32. Debrecen, Hungary, 1980.

[Mel82]   ———. An Alternative to Structure-Sharing. In K. L. Clark and S. A. Tarnlund, eds. *Logic Programming*, pp. 99–106. Academic Press, 1982.

[Mor78]   F. Morris. A Time and Space Efficient Garbage Compaction Algorithm. *CACM 21*, no. 8 (1978): 662–665.

[MT87]   H. Mulder and E. Tick. A Performance Comparison Between PLM and 68020 Prolog Processor. In J. L. Lassez, ed. *Fourth International Logic Programming Conference*, pp. 59–73. MIT Press, 1987.

[Nai85a]   L. Naish. Automating Control for Logic Programs. *Journal of Logic Programming* 2, no. 3 (1985):167–183.

[Nai85b]   ———. *Negation and Control in Prolog*. Ph.D. thesis, University of Melbourne, 1985.

[Nai86]   ———. Negation and Quantifiers in NU-Prolog. In E. Shapiro, ed. *Third International Logic Programming Conference*, pp. 624–634. LNCS 225. Springer-Verlag, 1986.

[Nai88]   ———. Parallelizing NU-Prolog. In R. A. Kowalski and K. A. Bowen, eds. *Fifth International Logic Programming Conference*, pp. 1546–1564. MIT Press, 1988.

[Nei79]   G. Neil. *Artificial Intelligence*. Tab Books, 1979.

[Nil84]   M. Nilsson. The World's Shortest Prolog Interpreter. In J. A. Campbell, ed. *Implementations of Prolog*, pp. 87–92. Ellis Horwood, 1984.

[NN87]   H. Nakashima and K. Nakajima. Hardware Architecture of the Sequential Inference Machine PSI-II. In *4th IEEE Symposium on Logic Programming*. San Francisco, 1987.

[NYY*83]   H. Nishikawa, M. Yokota, A. Yamamoto, K. Taki, and S. Uchida. The Personal Sequential Inference Machine (PSI). In *Logic Programming Workshop*, pp. 53–73. Algarve, Portugal, 1983.

[Per82]   J. F. Perrot. Le langage Prolog. Cours de DEA. Université Paris VI, 1982.

[Per84a]   F. Pereira. *C-Prolog Reference Manual*. D.A.I. University of Edinburgh, 1984.

[Per84b]   L. M. Pereira. Logic Control with Logic. In J. A. Campbell, ed. *Implementations of Prolog*, pp. 177–193. Ellis Horwood, 1984.

[Piq84]   J. F. Pique. Drawing Trees and Their Equations in Prolog. In *Second International Logic Programming Conference*, pp. 23–33. Uppsala, Sweden, 1984.

[Pla84]   D. Plaisted. The Occur Check Problem in Prolog. *New Generation Computing* 2 (1984): 309–322.

[PM83]   T. Pietrzykowski and S. Matwin. Intelligent Backtracking for Automated Deduction in FOL. In *Logic Programming Workshop*, pp. 186–191. Algarve, Portugal, 1983.

[PP82]   L. M. Pereira and A. Porto. Selective Backtracking. In K. L. Clark and S. A. Tarnlund, eds. *Logic Programming*, pp. 107–114. Academic Press, 1982.

[PPW78]   L. M. Pereira, F. Pereira, and D. H. Warren. *User's Guide to Dec-System 10 Prolog*. Technical Report. D.A.I. University of Edinburgh, 1978.

[Qui86]   *Quintus Prolog User's Guide*. Mountain View, Calif.: Quintus Computer Systems, 1986.

[RH86]   O. Ridoux and S. Le Huitouze. Une expérience du gel et du dif dans MALI. In *Journées Programmation en Logique du CNET*, pp. 269–280. 1986.

[RK85]   J. P. Roy and G. Kiremitdjian. *Lisp*. Cedic Nathan, 1985.

[Rob65]   J. A. Robinson. A Machine Oriented Logic Based on the Resolution Principle. *J.A.C.M.* 12, no. 1 (1965): 23–44.

[Rob71]   J. A. Robinson. Computational Logic, the Unification Computation. *Machine Intelligence* 6 (1971): 63–72.

[Rou75]   P. Roussel. *Prolog: manuel de référence et d'utilisation*. Technical Report. G.I.A. Université d'Aix-Marseille, September 1975.

[RS80a]   J. A. Robinson and E. E. Sibert. *Logic Programming in Lisp*. Technical Report 80-8. University of Syracuse, 1980.

[RS80b]   ———. *Loglisp: An Alternative to Prolog*. Technical Report 80–7. University of Syracuse, 1980.

[RS82]   ———. Loglisp: Motivation, Design and Implementation. In K. L. Clark and S. A. Tarnlund, eds. *Logic Programming*, pp. 299–313. Academic Press, 1982.

[RS87]   M. Ratcliffe and J. C. Syre. The PEPSys Parallel Logic Programming Language. In *IJCAI87*, pp. 48–55. 1987.

[Sha88]   E. Shapiro. *Concurrent Prolog*, vols. 1 and 2. MIT Press, 1988.

[SPP87]   *SP-Prolog Reference Manual*. BullSA, 1987.

[SS87]   E. Shapiro and L. Sterling. *The Art of Prolog*. 2d ed. MIT Press, 1987.

[Ste84]   G. L. Steele. *Common-Lisp: The Language*. Digital Press, 1984.

[TF83]   A. Takeuchi and K. Furukawa. Interprocess Communications in Concurrent Prolog. In *Logic Programming Workshop*, pp. 171–185. Algarve, Portugal, 1983.

[The88]   *The Arity/Prolog Language Reference Manual*. Concord, Mass.: Arity Corporation, 1988.

[Tic83]   E. Tick. *An Overlapped Prolog Processor*. Technical Report 308. SRI International, 1983.

[Tur86]   A. Turk. Compiler Optimizations for the WAM. In E. Shapiro, ed. *Third International Logic Programming Conference*, pp. 657–662. LNCS 225. Springer-Verlag, 1986.

[TZ86]   J. Thom and J. Zobel. *NU-Prolog Reference Manual, Version 1.0*. Melbourne University, 1986.

[Ued85]   K. Ueda. *Guarded Horn Clauses*. Technical Report TR-103. Tokyo: ICOT, 1985.

[Ued86]   ———. Making Exhaustive Search Programs Deterministic. In E. Shapiro, ed. *Third International Logic Programming Conference*, pp. 270–282. LNCS 225. Springer-Verlag, 1986.

[War77]   D. H. Warren. *Implementing Prolog: Compiling Predicate Logic Programs*. Technical Report 39–40. D.A.I. Edinburgh, 1977.

[War79]   D. H. Warren. Prolog on the DEC-System 10. In D. Michie, ed. *Expert Systems in the Micro-Electronic Age*. Edinburgh, 1979.

[War80]   ———. An Improved Prolog Implementation Which Optimizes Tail Recursion. In *Logic Programming Workshop '80*, pp. 1–11. Debrecen, Hungary, 1980.

[War83]   ———. *An Abstract Prolog Instruction Set*. Technical Report 309. SRI International, 1983.

[War87a] ———. *Or*-parallel Execution of Prolog. In LNCS 250, ed. *TAPSOFT*, pp. 243–259. 1987.

[War87b] ———. The SRI Model for *Or*-parallelism Execution of Prolog: Abstract Design and Implementation Issues. In *Fourth IEEE Symposium on Logic Programming*, pp. 92–102. 1987.

[WCS87]   A. Walker, M. Mac Cord, and J. F. Sowa. *Knowledge Systems and Prolog*. Addison Wesley, 1987.

[Win77]   P. H. Winston. *Artificial Intelligence*. Addison Wesley, 1977.

[Xil84]   *Xilog: manuel de référence*. BullSA, 1984.

Milton Keynes UK
Ingram Content Group UK Ltd.
UKHW022349220924
448641UK00006B/230